EAI/Springer Innovations in Communication and Computing

Series editor
Imrich Chlamtac, European Alliance for Innovation, Ghent, Belgium

Editor's Note

The impact of information technologies is creating a new world yet not fully understood. The extent and speed of economic, life style and social changes already perceived in everyday life is hard to estimate without understanding the technological driving forces behind it. This series presents contributed volumes featuring the latest research and development in the various information engineering technologies that play a key role in this process.

The range of topics, focusing primarily on communications and computing engineering include, but are not limited to, wireless networks; mobile communication; design and learning; gaming; interaction; e-health and pervasive healthcare; energy management; smart grids; internet of things; cognitive radio networks; computation; cloud computing; ubiquitous connectivity, and in mode general smart living, smart cities, Internet of Things and more. The series publishes a combination of expanded papers selected from hosted and sponsored European Alliance for Innovation (EAI) conferences that present cutting edge, global research as well as provide new perspectives on traditional related engineering fields. This content, complemented with open calls for contribution of book titles and individual chapters, together maintain Springer's and EAI's high standards of academic excellence. The audience for the books consists of researchers, industry professionals, advanced level students as well as practitioners in related fields of activity include information and communication specialists, security experts, economists, urban planners, doctors, and in general representatives in all those walks of life affected ad contributing to the information revolution.

About EAI

EAI is a grassroots member organization initiated through cooperation between businesses, public, private and government organizations to address the global challenges of Europe's future competitiveness and link the European Research community with its counterparts around the globe. EAI reaches out to hundreds of thousands of individual subscribers on all continents and collaborates with an institutional member base including Fortune 500 companies, government organizations, and educational institutions, provide a free research and innovation platform.

Through its open free membership model EAI promotes a new research and innovation culture based on collaboration, connectivity and recognition of excellence by community.

More information about this series at http://www.springer.com/series/15427

Sara Paiva • Suman Paul
Editors

Convergence of ICT and Smart Devices for Emerging Applications

 Springer

Editors
Sara Paiva
School of Technology and Management
Inst Politécnico de Viana do Castelo
Viana do Castelo, Portugal

Suman Paul
Department of ECE, School of Electronics,
Computer Science and Informatics
Haldia Institute of Technology
Haldia, West Bengal, India

ISSN 2522-8595　　　　　　　　ISSN 2522-8609　(electronic)
EAI/Springer Innovations in Communication and Computing
ISBN 978-3-030-41370-5　　　　ISBN 978-3-030-41368-2　(eBook)
https://doi.org/10.1007/978-3-030-41368-2

This Springer imprint is published by the registered company Springer Nature Switzerland AG.
The registered company address is: Gewerbestrasse 11, 6330 Cham, Switzerland

Preface

New information and communication technologies (ICT) have been developing at a very rapid pace, and are having an impact on various policy areas within new technologies, equipment, and smart devices. In particular, the use of these evolutions to improve the welfare of the population is notorious.

In this book, we intend to gather valuable contributions to help the reader understand the latest technological developments to help people with their various daily activities. In this context, this book includes several examples that use ICT and smart devices to target different domains of applications and segments of people.

The first two chapters introduce and contextualize the scope of this book by addressing smart technologies and their relevance to achieving sustainable goals as well as an analysis and classification of wearable device usability – equipment nowadays critical in developing ICT solutions. The two following chapters focus on the adoption of ICTs for specific segments of the population, namely, the ease of currency detection by the visually impaired and also a proposal for smart systems that help older people in their homes. Next is a chapter applied to agriculture and how ICT can help. Finally, the last two chapters address security aspects that should be considered in cloud environments and recommendation systems.

This book attracted contributors from several countries. Therefore, we would like to thank all authors for their contributions and also thank all reviewers for their review work. Last but not the least, we both express our sincerest thanks to Dr. Imrich Chlamtac, Series Editor, and Ms. Mary E. James, Senior Editor, EAI/Springer. We convey special thanks to Ms. Eliška Vlčková, Managing Editor, European Alliance for Innovation (EAI), for all her support and cooperation throughout the process of creating this book.

Viana do Castelo, Portugal   Sara Paiva
Haldia, India   Suman Paul

Contents

Convergence of Open Data, Digital Libraries, and Smart Technologies for Accelerating Progress Toward Sustainable Development Goals

Ibrahim Sidi Zakari

1 Introduction

The convergence of information and communication technologies is gaining momentum as highlighted by several authors including Huang et al. [11].

During the last two to three decades, the boundaries between information technology (IT), which refers to hardware and software used to store, retrieve, and process data and communications technology (CT), which includes electronic systems used for communication between individuals or groups, have become increasingly indistinguishable.

For example, at the end of the 1990s, people used to gather in large computing rooms for data processing purposes. Computing tasks are currently performed thoughout networks of geographically separated but interconnected devices. The use of massive datasets and advanced simulations and computation methods is impacting positively scientific research and knowledge co-creation.

Moreover, in many places around the world, mobile Internet users (e.g., 3G and 4G, or third-generation and fourth-generation, mobile phone subscribers) outnumber fixed Internet users (e.g., personal computer users).

The convergence of information and communication technologies (ICT) and smart devices is contributing to the production and dissemination of various types of data including administrative data, commercially licensed data, geospatial data, metadata (e.g., call detail records), official statistics, open data, photo or video data, private sector data, qualitative data, satellite imagery, sensor data, survey data, unstructured data, web or social media data and paradata (e.g., geographic location of the respondents, used devices, browsers, and platforms; level of vocabulary in a

I. Sidi Zakari (✉)
Department of Mathematics and Computer Science, Abdou Moumouni University of Niamey, Niamey, Niger

© Springer Nature Switzerland AG 2020
S. Paiva, S. Paul (eds.), *Convergence of ICT and Smart Devices for Emerging Applications*, EAI/Springer Innovations in Communication and Computing, https://doi.org/10.1007/978-3-030-41368-2_1

text). Facts that are fostering synergies between communities of non-official data (e.g. Academia, Industry and Open Street Map communities) and official statistics producers (e.g. National Statistical Offices).

Furthermore, smartphones and tablets combine computer, maps, TV, GPS, telephone, camera, projector, alarm clock, personal research assistant, music player, newspaper, translator, flashlight, web browsing, data sharing, and data storage by providing vast storage capacity and acting as a modem for Internet access to other devices.

Moreover, we are currently living in the era of the Internet of Things (IoT) where physical devices have the ability to collect data, exchange information, make decisions, and control themselves or other devices. Combinations of statistical analysis, semantic technology, personal assistants, and advanced irtificial intelligence are increasing human–machine interactions. The book[1] entitled *Internet of Things: Converging Technologies for Smart Environments and Integrated Ecosystems* (2013) provided an overview of various topics of the IoT including the R&D priorities. Nowadays, many governments around the world are harnessing the potential of ICT to build a more transparent, efficient, and inclusive relationship with citizens. This new paradigm of open government (introduced in 2007), which is interrelated with e-government, differs among countries due to many reasons including technological and socioeconomic ones.

E-government systems allow citizens to receive government information, provide feedback, and carry out needed transactions in real time. Fixed and portable/wearable sensors and mesh networks are making cities and villages smarter, able to diagnose and fix local problems.

The concept of open data was introduced in the middle of year 2000, and which is closely related to open government, represents a more proactive data-based communication and interaction with citizens and end users as well as an opportunity of value-added services for private sector. The state of open data has been recently investigated by Davies et al. [8].

Countries are updating their national strategies for ICT by adapting the corresponding institutional, legal, and regulatory frameworks related to electronic communication for taking into account technological evolution, the convergences of networks and systems, and information society requirements.

ICT are fundamental for accelerating the achievement of the 17 United Nations Sustainable Development Goals (SDGs) to transform our world. Wu et al. [19] recently investigated the landscape of ICT for SDGs with a focus on state-of-the-art needs and perspectives, while Kostoska and Kocarev [12] proposed a novel ICT framework for SDGs.

This chapter is a contribution for advancing knowledge related to the convergence of open data portals, digital libraries, interactive whiteboards, and smart devices widespread that conduct to emerging applications and data innovations related to the United Nations Sustainable Development Goals (SDGs).

---

[1]https://www.riverpublishers.com/book_details.php?book_id=176

The rest of the chapter is organized as follows: the first section dedicated to general concepts and terminologies is followed by the research methodology, the analysis of some illustrative case studies, and the concluding remarks.

2 General Concepts and Terminologies

The concept of convergence has many definitions as highlighted in the lines below.

2.1 What Is Convergence?

Convergence is the coming together of two different entities, and in the contexts of computing and technology, is the integration of two or more different technologies in a single device or system. A good example is the convergence of communication and imaging technologies on a mobile device designed to make calls and take pictures - two unrelated technologies that converge on a single device.[2]

Convergence is a deep integration of knowledge, tools, and all relevant activities of human activity for a common goal, to allow society to answer new questions to change the respective physical or social ecosystem. Such changes in the respective ecosystem open new trends, pathways, and opportunities in the following divergent phase of the process [3, 16].

2.2 What Is Technology Convergence and How Is It Possible?[3]

A convergence is when two or more distinct things come together. Technology convergence is when different forms of technologies cohabitate in a single device, sharing resources and interacting, creating new technology and convenience.

Borés et al. [6] define technological convergence as "a process by which the telecommunications, broadcasting, information technologies and entertainment sectors (collectively known as ICT – Information and Communications Technologies) may be converging towards a unified market." Technological convergence has both technical and functional sides. The technical side refers to the ability of any infrastructure to transport any type of data, whereas the functional side means the ability of consumers to seamlessly integrate various functions (computation, entertainment and voice) in unique devices.

---

[2]https://www.techopedia.com/definition/769/convergence
[3]https://shape.att.com/blog/technology-convergence

2.3   Examples of Convergence

Digital convergence occurs when the same multimedia content can be displayed on different types of devices (due to digitilization), and this binary information can be stored, published, and sent in an efficient manner. This is not possible for analog data and other data types, which are modified during the copy or transmission processes. This digital convergence is making technological convergence possible.

Technological Convergence in Education (EdTech)
Technology has brought about a revolution in the way knowledge is passed from a teacher to a student. Various devices such as smartphones and tablets are being introduced to work in conjunction with cloud-based models at the back end for the easy sharing of content among knowledge seekers. Numerous platforms are developed by universities for open discussions among researchers to better decompose problems and identify more optimal solutions.

ICT can contribute to quality of education and universal access to lifelong learning opportunities.

Despite the ability of technological advances to improve quality of education, some barriers reduce the effectiveness of the use of ICT in the learning environment (Blackwell et al. [5]).

Graham and Michael [9] and Annika and Åke [2] also highlighted the importance of ICT integration when teaching literacy and mathematics.

Although M-Learning is contributing to track students' progress in many areas (thanks to mobile phone widespread use), smartphone penetration rate (by 2017) was only expected to reach 20% (Agence Française de Développement [1]).

2.4   Technology Convergence Regulatory Issues

Access to intellectual property (IP) represents one of the biggest challenges in the multimedia industry and regulation of digital data. However, solutions for preventing copyright crimes include trademarking and copyrighting of creative contents through mechanisms like the Digital Millennium Copyright Act (DMCA).[4]

2.5   Open Data

The relationships between different types of data were highlighted in Guyon et al. [10] through Fig. 1.

---

[4]https://www.copyright.gov/legislation/dmca.pdf

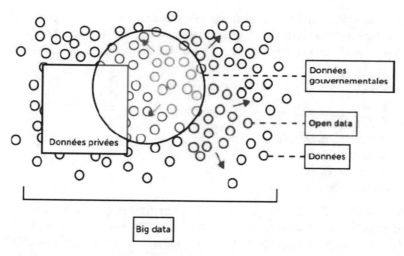

Fig. 1 The relationships between different types (Données = data; Données gouvernemen-tales = Government data; Données privées = private data) of data © Sophie Czich (CC) BY-NC-SA

Open data are public or third-party (private sector, NGO, research institutions, etc.) data offered to users according to certain conditions.

Private data (in the sense of personal data) are data protected by the laws on the protection of personal data and do not enter by principle in the world of open data. Government data is potentially public data diffusable under the name of open data.

Other data are those held by companies or individuals which can potentially also be offered as open data. Big data is all big data managed publicly or privately (in the sense of third party), with open data included, but also a multitude of micro-data collected by all applications primarily on the Internet.

These conditions were initially defined during a meeting in Sebastopol (California), then completed in the article published by the Sunlight Foundation "Ten Principles for Open Government Information," that is, "ten principles for the opening of government information."[5] Open data is the idea that certain data should be freely available to everyone to use and republish as they wish, without restrictions from copyright or any other restrictions (Ubaldi [18]).

2.6 Open Government Data[6]

According to the Open Government Data (OGD) website, "open" means data is open, i.e., "free for anyone to use, re-use, and re-distribute," and "Open Government

---

[5]https://sunlightfoundation.com/policy/documents/ten-open-data-principles/
[6]Open Government Data. (2015). Retrieved July 10, 2019 from http://opengovernmentdata.org/about/

Data" means "data and information produced or commissioned by government or government controlled entities." The government data shall be considered open if it is publicly available in a way that complies with the following eight principles[7]:

   (i) *Complete*
  (ii) *Primary*
 (iii) *Timely*
 (iv) *Accessible*
  (v) *Machine processable*
 (vi) *Nondiscriminatory*
 (vii) *Nonproprietary*
(viii) *License-free.*

According to Ubaldi [18], "a number of challenges may be associated with the implementation of OGD initiatives which, if not properly tackled, might obstruct or restrict the capture of benefits of national efforts aimed at spurring OGD." The challenges are:

(a) Harmonization of government data (multiple sources of data, different formats, and standards).
(b) Interoperability.

The lack in guidelines for regulating and helping in the process of opening data (for transparency and accountability of governments) has been raised by Nugroho [15]. According to Braunschweig et al. [7], availability of the data online is not sufficient; and some legal, administrative, and technical requirements need to be fulfilled by publication platforms.

3   Research Methodology

We investigate the convergence of ICT and smart devices through the following methodology which is based on our research on UN SDGs or the implementation of open data projects and illustrative case studies from the integration of ICT in education:

1. Understanding the linkages between ICT and the SDGs.
2. Analysis of digital convergence through an experimental open data portal for promoting data innovations and visualizations related to SDGs.
3. Analysis of digital and technological convergences in the context of integration of digital libraries for education in areas without the Internet.
4. Analysis of digital and technological convergences in the context of the integration of interactive whiteboard and smart microprojector for teaching statistics.

---

[7]Open Government Data Principles. (2015). Retrieved July 10, 2019 from https://public.resource. org/8_principles.html

The following sections will comment in more detail on these four different strands to the research.

3.1 ICT and the UN Sustainable Development Goals (SDGs)

Huawei's 2019 ICT Sustainable Development Goals Benchmark report[8] highlighted (Fig. 2) the fact that ICT can make the most difference in achieving SDG #4 (Quality Education), SDG #3 (Good Health and Well-being), SDG #9 (Industry, Innovation, and Infrastructure), SDG #5 (Gender Equality), SDG #11 (Sustainable Cities and Communities), and SDG #7 (Affordable and Clean Energy).

For example in the case of SDG #4 (*72% correlation with ICT*), the report stated that "In the future, all development is inseparable from talents and culture. However, there are currently more than 265 million children out of school worldwide, which greatly hinders social and economic development. The high correlation between SDG 4 and ICT skills indicates that a country's overall education level is closely related to its ICT education and training level. Only when the need for the required skills is satisfied can we promote the fair development of the whole society."

In the case of SDG #9 (*63% correlation with ICT*) it is mentioned that "16% of the global population does not have access to mobile broadband networks. Research has found that people and organizations' ability to access and use ICT

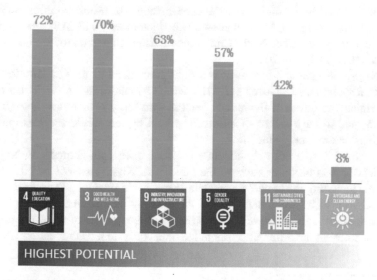

Fig. 2 Highest correlations between SDGs and ICT. (Source: 2019 ICT Sustainable Development Goals Benchmark report)

---

[8]https://www.huawei.com/en/about-huawei/sustainability/sdg, accessed on 22 July 2019.

services is more likely to drive economic development than ICT education and skills. Therefore, countries should focus on improving the access and use of ICTs, increase investment in industry, innovation and infrastructure, promote industrial development, and enable people to enjoy the social and economic dividends brought about by ICT development and promote social equity development."

The previous report also identified three critical pathways for ICT to drive sustainable development: (a) increase access to information and services, (b) increase connectivity between people and organizations, and (c) increase productivity and resource efficiency.

Furthermore, as part of the data for climate action challenge,[9] organized in 2017, leading companies around the world provided data to researchers and cloud computing support from Microsoft and/or data visualization support from Tableau. These datasets anonymized to protect privacy allowed participants to generate new solutions to help meet Sustainable Development Goal #13: Climate Action. However, because the 2030 Agenda is integrated and indivisible and climate action affects the attainment of its other goals, the challenge was also designed to support achievement of the 2030 Agenda as a whole, including a specific thematic focus on how climate action relates to the other goals.

3.2 Digital Convergence and Open Data Portals

This section highlights digital convergence through an award-winning open data portal initiative (Fig. 3), implemented since September 2018 in Niger as part of strategic projects identified by the Conference Afrique Francophone sur les Données Ouvertes (CAFDO)[10] in 2017.

In Niger, the right to information is a right guaranteed by the Constitution of the Seventh Republic on 25 November 2010, which stipulates in Article 31 (in respect of the rights and duties of the human person) that "Any person has the right to be informed and to have access to information held by the public services under the conditions determined by the law."

This portal[11] will not only federate existing open data sources on Niger but also identify data needs to improve the ranking of Niger vis-à-vis of international standards including the Global Open Data Index[12] and the Open Data Barometer.[13]

It should also be noted that most of the existing data sources are the result of isolated initiatives that have not necessarily been the subject of wide-ranging consultation between the different actors in the data ecosystem (public administrations,

---

[9]http://dataforclimateaction.org/

[10]https://www.donneesouvertes.africa

[11]https://odn.datafordev.org/

[12]https://index.okfn.org

[13]https://opendatabarometer.org

Fig. 3 Overview of the experimental open data portal (https://odn.datafordev.org/) highlighting data per categories as well as data analytics tools for descriptive statistics

sector private sector, civil society, technical and financial partners, universities and research institutes, etc.).

On the other hand, achieving the Sustainable Development Goals (SDGs) necessarily means producing (in a participatory manner) access (fair and at a lower cost) and using (by all) quality and reliable data so that no one will be left behind by 2030.

Indeed, in the age of the knowledge economy and SDGs, it is also important to educate citizens about the importance of open data for:

– Enjoyment of fundamental human rights
– Better decision-making with a view to improving their living conditions
– Promotion of quality services
– The follow-up of the action of the executive, legislative, and judicial power
– The promotion of good governance and the consolidation of democracy

Finally, Niger's accession to the Open Government Partnership (OGP) initiative is still under discussion, and we hope to make the necessary advocacy to speed up the process.

This experimental open data portal was developed under the CKAN platform, and its key functional features include the publication of data, exploring datasets, visualization of datasets, and access to data through an API. The interface was customized based on the portal's graphic charter and optimized for being responsive on various devices (Fig. 4).

In terms of data format, the datasets from this portal can be represented in a variety of formats such as PDF, XLS, CSV, JSON, RDF, LOD, and more. However, the standard formats for open data are nonproprietary formats such as CSV, JSON, and RDF. These formats enable the users to further reuse the datasets without having to purchase software to process them. Formats such as the LOD promote the concept of open linked data, which is the apex of the five-star rating.[14]

Finally, for datasets published on the data portal to be reusable by users, it has to be published under an open data license. The most recognized and commonly used data licenses are Creative Commons (CC) licenses, CC zero (CC0), Open Data Commons, and Open Government License.

Digital Content Diffusion Solutions in Area without the Internet and Library Network Collective Catalog[15]: The Case of the Culture Box

The CultureBox (Fig. 5) is an innovative project conceived and launched by the media library of Franco-Nigerien Cultural Center (CCFN) of Niamey (Niger), with the support of the Institut Français de Paris. The French company Mind & Go ensures its development. In a context of digital divide, it is a mobile, autonomous

---

[14]https://5stardata.info/en/

[15]https://mediatheques-niger.org/index.php?lvl=index

Fig. 4 Digital convergence of the experimental open data portal on different devices

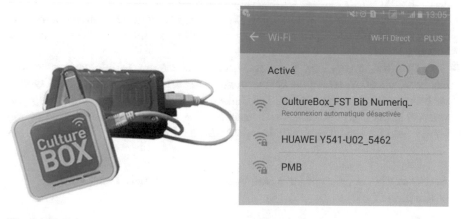

Fig. 5 The CultureBox digital content library accessible via Wi-Fi for areas without the Internet

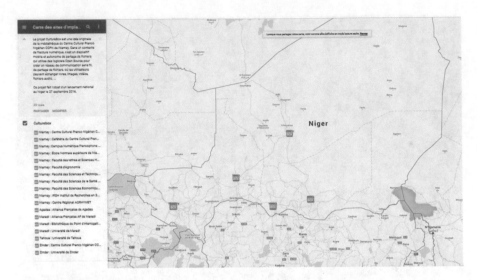

Fig. 6 Mapping of the CultureBox network in Niger © Éric Durel

file-sharing device that uses open-source software to create a wireless file-sharing communication network, where users can exchange books, images, videos, audio files, and more. CultureBox is a new way to make documents available in libraries, with a philosophy of free sharing of common goods that symbolizes the public domain.

The CultureBox was launched in Niger on September 27, 2016, and it is no less than 22 Culture Boxes that were deployed on 19 sites in Niger (see Figs. 6 and 7 and Table 1).

Those Boxes constitute an Offline Digital Library with 22 stand-alone routers deploying an offline digital file-sharing network in as many libraries across the country.

CultureBox is very easy to access. It does not require an Internet connection: just be within 50 meters of the device, connect to cultureBox by Wi-Fi, then type in the browser www.culturebox.lan. The CultureBox page will open automatically, and you will have immediate access to thousands of resources (books, movies, speeches, games, etc.) free and downloadable in less than a minute.

Access to Digital Content for Areas Without the Internet: The Case of the DataCup[16]

Accessible from smartphones, tablets, computers, or even televisions, DataCup (Figs. 8 and 9) offers access to thousands of educational and cultural resources.

---

[16]https://datacup.io/-La-DataCup-

Fig. 7 Overview of the CultureBox network and integrated multimedia content

Table 1 Places with CultureBox in Niger

Sites	ID	Region
Centre Culturel Franco Nigérien CCFN de Niamey	CCFN Niamey	Niamey
Cafétéria du Centre Culturel Franco Nigérien CCFN de Niamey	Cafeteria CCFN Niamey	Niamey
UAM-Bibliothèque Universitaire Centrale	BUC Niamey	Niamey
UAM-Campus Numérique Francophone CNF de Niamey	CNF	Niamey
UAM- École Normale supérieure de Niamey	ENS	Niamey
UAM-Faculté des lettres et Sciences Humaines (FLSH)	FLSH	Niamey
UAM-Faculté d'Agronomie	Agro	Niamey
UAM-Faculté des Sciences et Techniques (FST)	FST	Niamey
UAM-Faculté des Sciences de la Santé (FSS)	FSS	Niamey
UAM-Faculté des Sciences Economiques et juridiques (FSEJ)	FSEJ	Niamey
UAM-IRSH Institut de Recherches en Sciences Humaines	IRSH	Niamey
Centre Régional AGRHYMET	AGRHYMET	Niamey
Alliance Française d'Agadez	AF Agadez	Agadez
Alliance Française AF de Maradi	AF Maradi	Maradi
Bibliothèque du Point d'interrogation BPI	BPI Maradi	Maradi
Université de Maradi	BUC Maradi	Maradi
Université de Tahoua	BUC Tahoua	Tahoua
Centre Culturel Franco Nigérien CCFN de Zinder	CCFN Zinder	Zinder
Université de Zinder	BUC Zinder	Zinder

Fig. 8 Digital convergence through the DataCup deployed at Abdou Moumouni University of Niamey

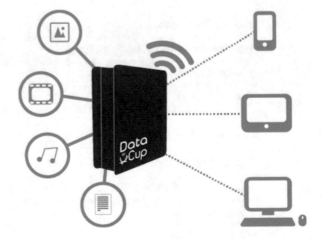

Fig. 9 Design and hardware aspects of the DataCup

Deployable in any public place, it offers access to royalty-free, interoperable, public domain, or Creative Commons resources that can be consulted and downloaded: encyclopedias, books, tales and audiobooks, podcasts, photo library, video library, scores, tutorials, theses, games, etc.

No need for specific software, a simple browser is enough; additional features include a powerful connectivity and high number of simultaneous consultations through all existing media.

Powered by Mind And Go, a specialist in OpenSource management solutions including library management solutions, the DataCup comes from a UNHCR's request for cultural animation in refugee camps in Niger.

On the bases of concrete needs users faced with problematic of stability of electricity and Internet networks in addition to the issues like so-called tropical climate, Mind And Go has designed a simple solution, reliable and robust, distributed in Open Hardware. Based on standard market components and accessible materials, DataCup resists to intense use; it can be transported easily and repaired anywhere.

DataCup is thought to distribute documents offline (more than 250GB royalty-free data) on all types of media available for users (a Wi-Fi access point is accessible on tablets, smartphones, and computers for about 150 users).

Scheduled to be updated in terms of content, DataCup automatically connects itself to a centralized platform available in France and offeris a common varied catalog with many accessible documents.

Since 2017 and after 2 years of research and development, DataCup is present in the camps of the UNHCR of Niger, within the antenna Humanity and Inclusion (formerly Handicap International) in Niamey, and 14 prototypes are currently being deployed at Abdou Moumouni University (UAM) in Niamey.

Refugees from UNHCR can thus discover documentaries, referring to their place of origin just as the Abdou Moumouni University students will soon be able to consult all the digitized theses of all their faculties, and this for free.

Coming soon, e-learning or mapping solutions will be easily deployable.

3.3 Digital and Technological Convergences in the Context of the Integration of Interactive Whiteboard and Smart Microprojector for Statistical Education

In the award-winning video[17] on innovation in pedagogy entitled *Ressources Numériques Pédagogiques et Apprentissage de la Statistique*, Sidi Zakari illustrated some of his innovations in the field of statistics education (bachelor level at Abdou Moumouni University in Niamey, Niger) through:

1. The use of the culture box digital library (previously discussed)
2. The Use of an interactive whiteboard
3. The use of brainstorming
4. The use of a smart microprojector with electric autonomy (also integrating multimedia, USB and HDMI ports, search engines, QR code, Wi-FI, and Bluetooth technologies)

Marzano and Haystead [13] indicated that the usage of interactive whiteboard has risen student achievement by 16%.

Becta defined interactive whiteboard as "a large, touch-sensitive board, which is connected to a digital projector and a computer. The projector displays the image from the computer screen on the board. The computer can then be controlled by touching the board, either directly or with a special pen" [4].

According to Moss et al. [14], "Combining touch sensitive screens with digital projection opens up new possibilities. In terms of audience presentation, the combination of digital projection and a touch sensitive screen allows the presenter to operate from the screen itself without having to go to the computer. Using a hand or pen on the screen like a mouse, the user can then move about within that environment with exactly the same kind of functionality associated with mouse use at a computer terminal: clicking, dropping and dragging, or scrolling. This makes it possible to exploit different kinds of computer software and the choices they offer whilst any presentation is in process, including making use of the internet by moving around and between websites; as well as using the full potential of the tool bar and its menus to zoom in and out on images, cut and paste within documents and open up new windows. In this way, new texts can be created from the board as the display proceeds."

Smart Technologies Incorporation [17] listed several functions of Interactive WhiteBoard in its White Paper as follows (Fig. 10).

Currently interactive whiteboards integrate simultaneous touch differentiation function for multiuser collaboration which means many students can work simultaneously.

---

[17]First prize of the Agence Universitaire de la Francophonie's competition 2018 on innovation in pedagogy entitled « Mon innovation en 120 secondes » available on Youtube via https://youtu.be/v8uSGVz1-80

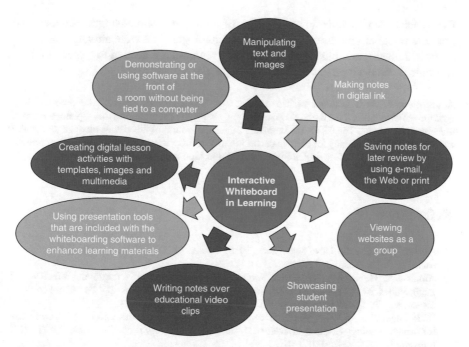

Fig. 10 Several functions of Interactive WhiteBoard. (Source: https://sites.google.com/a/myport. ac.uk/ict-and-the-secondary-classroom-esliarozhi/project-definition/interactive-white-board)

4 Concluding Remarks

In this chapter, we investigated recent advanced topics related to the convergence of open data portals, digital libraries, interactive whiteboards, and smart devices widespread that conduct to emerging applications and data innovations related to the United Nations Sustainable Development Goals (SDGs).

Technological convergence plays a crucial role in society from sustainable development perspectives and for bridging the digital divide and data gap.

Digital technologies provide solutions for more efficient ways to collect and analyze large sets of data with the help of big data analytical tools, which have wide-ranging implications for SDG's progress.

Technological convergence, together with technological standardization, can enable transparent and modular communication between diverse devices over the network and provides advantages for service providers for coordinated and more efficient service delivery. However, technological convergence has some limitations related to interoperability, interconnection, consumer protection, and universal access. Others include new regulatory frameworks related to intellectual property, licensing and regulation of providers, bandwidth shortage, infrastructure upgrades, strategic alignment by service providers, privacy, security, and reliability.

Drawbacks also include lower quality of multiple task-converged devices and possible waste of investments in separate technologies that were already made.

In this regard, continuous effort and collaboration are needed to better facilitate technological development through convergence.

Acknowledgments The author wishes to thank Conférence Afrique Francophone pour les Données Ouvertes (CAFDO) for the small grant and the technical support during the implementation of our open data project. Thanks also to Bemalera Toussaint Koura for the visuals on open data portal and Mind And Go company for the wonderful job and collaboration during the implementation of the CultureBox and DataCup digital libraries at Abdou Moumouni University of Niamey.

References

1. Agence Française de Développement. (2015, February). *Digital services for education in Africa* (PDF).unesco.org. Retrieved 23 July 2019.
2. Annika, A. G., & Åke, G. (2016, August). Closing the gaps – Improving literacy and mathematics by ict-enhanced collaboration, Science Direct, 2016, pg 78. *Computers & Education, 99*, 68–80. https://doi.org/10.1016/j.compedu.2016.04.004.
3. Bainbridge, W. S., & Roco, M. C. (2016). Science and technology convergence: With emphasis for nanotechnology-inspired convergence. *Journal of Nanoparticle Research, 18*(7), 211. https://doi.org/10.1007/s11051-016-3520-0.
4. Becta. (2003). *What the research says about interactive whiteboards*. http://www.becta.org.uk/page_documents/research/wtrs_whiteboards.pdf
5. Blackwell, C. K., Lauricella, A. R., & Wartella, E. (2014). Factors influencing digital technology use in early childhood education. *Computers & Education, 77*, 82–90.
6. Borés, C., Saurina, C., & Torres, R. (2003). Technological convergence: A strategic perspective. *Technovation, 23*(1), 1–13.
7. Braunschweig, K., et al. (2012). *The state of open data: Limits of current open data platforms*. Retrieved from http://www2012.org/proceedings/nocompanion/wwwwebsci2012_braunschweig.pdf
8. Davies, T., Walker, S., Rubinstein, M., & Perini, F. (Eds.). (2019). *The state of open data: Histories and horizons*. Cape Town and Ottawa: African Minds and International Development Research Centre. First published in 2019 by African Minds and the International Development Research Centre (IDRC).
9. Graham, S., & Michael, H. (2010). *Writing to read: Evidence for how writing can improve. Carnegie corporation time to act report*. Washington, DC: Alliance for Excellent Education. Print.
10. Guyon, C., Longin, C., & Zeller, J.-D. (2015). Open data et archivage électronique, quelles convergences? In *La Gazette des archives, n°240, 2015–4. Voyages extraordinairement numériques: 10 ans d'archivage électronique, et demain?* (pp. 385–396). https://doi.org/10.3406/gazar.2015.5320. http://www.persee.fr/doc/gazar_0016-5522_2015_num_240_4_5320.
11. Huang, I., Guo, R., Xie, H., & Wu, Z. (2012). Chapter 1.2: The convergence of information and communication technologies gains momentum. In: *The global information technology report 2012* (pp 35–45). The World Economic Forum. http://www3.weforum.org/docs/GITR/2012/GITR_Chapter1.2_2012.pdf
12. Kostoska, O., & Kocarev, L. (2019). A novel ICT framework for sustainable development goals. *Sustainability, 11*(7). ISSN 2071-1050.
13. Marzano, R. J., & Haystead, M. (2009). *Final report on the evaluation of the promethean technology*. Englewood, CO: Marzano Research Laboratory.

14. Moss, G., Jewitt, C., Levaãiç, R., Armstrong, V., Cardini, A., & Castle, F. (2007). *The Interactive Whiteboard, Pedagogy and Pupil Performance Evaluation: An Evaluation of the Schools Whiteboard Expansion (SWE) Project: London Challenge*. Research Report 816. Department for Education and Skills/Institute of Education, University of London.
15. Nugroho, R. (2013). A *comparison of open data policies in different countries: Lessons learned for an open data policy in Indonesia*. Retrieved from http://repository.tudelft.nl/view/ir/uuid%3Aae4e0a64-579d-40c4-bed0-d51614ddea9c/
16. Roco, M. C. (2002). Coherence and divergence of megatrends in science and engineering. *Journal of Nanoparticle Research, 4*, 9–19. https://doi.org/10.1023/A:1020157027792.
17. SMART Technologies Inc. (2006). *Interactive whiteboards and learning*. White Paper http://downloads01.smarttech.com/media/education/pdf/interactivewhiteboardsandlearning.pdf
18. Ubaldi, B. (2013). Open government data: Towards empirical analysis of open government data initiatives. *OECD Working Papers on Public Governance*, (22), 0_1.
19. Wu, J., Guo, S., Huang, H., Liu, W., & Xiang, Y. (2018). Information and communications technologies for sustainable development goals: State-of-the-art, needs and perspectives. *IEEE Communications Surveys & Tutorials, 20*(3), 2389–2406. https://doi.org/10.1109/COMST.2018.2812301.

A Comprehensive Framework
of Usability Issues Related
to the Wearable Devices

Jayden Khakurel, Jari Porras, Helinä Melkas, and Bo Fu

1 Introduction

Continual innovation in hardware and software technologies, such as sensors, displays, processors, storage memory, and algorithms, has been crucial in changing the paradigm of computing devices. Mobile computing has advanced rapidly over the past decade, and the components found in such computing devices are becoming increasingly smaller while remaining extremely powerful. The emergence of quantified-self technologies, including wearable devices, is one of the most evident examples of this technological development.

Wearable devices can be defined as, "smart electronic devices available in various forms; located near or on the human body to sense and analyze physiological and psychological data such as feelings, movements, heart rate, blood pressure, and so forth, via applications either installed on the device itself or on an external device (i.e., smartphones that are connected to the cloud)" (p.2) [1]. According to Motti and Caine [2], "since the first sensors were produced, the wearable device field has evolved exponentially" and "is characterized by body-worn devices, such as clothing and accessories" (p.1820). Humans use wearable devices in their daily

J. Khakurel (✉)
Research Center for Child Psychiatry, University of Turku, Turku, Finland
e-mail: jayden.khakurel@utu.fi

J. Porras · H. Melkas
LUT University, Lappeenranta, Finland
e-mail: jari.porras@lut.fi; helina.melkas@lut.fi

B. Fu
California State University, Long Beach, CA, USA
e-mail: bo.fu@csulb.edu

© Springer Nature Switzerland AG 2020
S. Paiva, S. Paul (eds.), *Convergence of ICT and Smart Devices for Emerging Applications*, EAI/Springer Innovations in Communication and Computing,
https://doi.org/10.1007/978-3-030-41368-2_2

21

activities to gather and assess a diverse range of data "from internal states (as mood or glucose level in the blood) to performance values (as pace or kilometers run), from habits (as food, sleep) to actions (as visited places)" (p.1) [3]. Lee et al. [4] note that in many applications areas ("i.e. areas of wellness, healthcare, assistance for the visually impaired, disaster relief, and public safety" (p.15), the development of wearable devices has contributed significantly to enhancing the quality of daily life of both individuals and society as a whole. It is expected that in the upcoming year, virtual reality (VR) headsets, such as Samsung VR, will be used as an alternative to conventional televisions, and Microsoft HoloLens and similar devices will enhance human vision. In addition, it is anticipated that smartwatches and mobile devices will assist users with health monitoring, for example, by making it possible for patients to monitor a bacterial infection or their glucose levels. In particular, wearable devices are more and more being seen as integral to a future in which users will control devices remotely via the Internet.

Despite the potential benefits of wearable device usage, numerous researchers have generally recognized that wearable technologies are failing to inspire long-term adoption [5–7]. For example, Lazar et al. [6] find that their participants abandoned almost 80% of their purchased wearable devices within the first 2 months because of deficiencies in usability. Another study by Endeavor Partners [7] reports that many wearable device users abandoned their devices within 6 months of initial usage because of poor experiences. Clawson et al. [5] indicate that individuals abandoned their wearable devices because (i) they were too complicated to use; (ii) they were too complex to learn; or (iii) they failed to help the users achieve their goals. Although the principal objective of these wearable devices is to provide the user with higher levels of ease and flexibility [8] in data acquisition without any degree of intrusiveness [9], usability is seen as one of the more influential factors associated with device abundancy. Furthermore, Piwek et al. [10] state that "wearable devices don't add functional value that is already expected from personal technology of that type, and they require too much effort, which breaks the seamless user experience" (p.3). Moreover, Motti and Caine [2] assert that "by focusing on the feasibility of an individual approach, often usability and wearability are neglected" (p.1821) on wearable devices. As asserted by Abbas [11], "The outcome of good usability is a greater likelihood of user acceptance. User acceptance is often the difference between a product's success or failure in the marketplace" (p.1764). Trivedi [12] also states, "The user is concentrating on the usability of the device. Therefore, usability has become an important parameter today" (p.69).

The term "usability" is derived from ISO standard 9241-11, where usability is described as the "extent to which a product can be used by specified users to achieve specified goals with effectiveness, efficiency and satisfaction in a specified context of use" [13]. Usability can also be construed as the value that users derive from using the technology or device. Gafni [14] states, "Usability is one of the most important characteristics when targeting systems to wide audiences that need to operate an intuitive system without direct training and support" (p.755). However, inappropriate design, lack of context-awareness will affect the usability while interacting with devices and interrupt individual to accomplish their goals [15, 16].

Therefore, usability parameters are extremely important to the success of wearable devices because they enable users to derive the full benefits of the device without requiring specific training or additional guidance [11, 12, 16, 17].

We argue that to improve the usability of wearable devices and increase their long-term use, it is first necessary to identify and then understand usability-related issues, especially regarding in which wearable device categories previous research has addressed these issues. Thus, the current study undertakes a systematic literature review (SLR) that follows the method presented by Petersen et al. [18]. An SLR presents an opportunity to closely review the current state-of-the-art [19] by synthesizing evidence to signify critical implications [20], identify unresolved problems, discover research trends, and create a basis for novel intervention.

The present work seeks to identify, understand, evaluate, and synthesize usability issues in the wearable device's domain including which usability evaluation methodology has been applied by the researchers. The time span considered begins in 2000, when based on Google trends and Motti and Caine [2] report, wearable devices were first introduced and marketed,[1] and ends in January 2018.

The rest of the paper is organized as follows: Sect. 2 presents a brief synopsis of the motivation and related work; Sect. 3 discusses how the research process was carried out to provide definitive results for the research questions (RQs); Sect. 4 presents the findings and interprets the results; Sect. 5 discusses the significance of the results and presents the limitations of the study; and Sect. 6 concludes the work by restating the main points made in the study and indicating future areas of study.

2   Related Work

As the field of mobile technology has advanced, wearable sensors that collect data from human activity have emerged [21]. Because these wearable devices are completely different from mobile devices in terms of their size, functionality, user interaction, and platform, their integration into people's daily lives poses a variety of challenges [22]. Finding the right balance between attributes such as "accessibility, usability, and wearability" [23] in wearable device remains difficult. One of the difficulties stems from the unique interaction modalities of these devices compared with other computing devices, especially in terms of the input-output mechanisms, which require a new design approach [2]. Because wearable devices are a comparatively new field of study, there are inadequate number of studies that have reviewed and analyzed usability and its relation to different types of wearable devices. For example, Motti and Caine [2] literature review identifies wearability principles. The study considers device characteristics, that is, hardware and software, arguing that 20 human-centered principles could help designers

---

[1]The history of wearable technology – Past, present and future. https://wtvox.com/featured-news/history-of-wearable-technology-2/

understand the design process and facilitate design with the focus on "users' wishes, interests, and requirements." The study concludes that even though these principles could overcome some obstacles, helping designers focus more on design than human factors when developing novel wearables, trade-offs such as technical and ergonomics requirements still require careful analysis. Similarly, Dhawale and Wellington [24] use an ethnographic study to identify the usability characteristics of activity monitoring devices and how these characteristics encourage the prolonged use of such devices. They focus on identifying the usability issues of these particular monitoring devices versus a larger sample with multiple wearable devices. The study identifies six key usability characteristics that play a vital role for device users: "display size of the screen, weight of the device, battery life, multitasking, social engagement, and ease of use." Other authors, for example, Jiang et al. [25] reviewed how and why wearable devices are developed and why they have gained popularity in recent years. Their work includes a consideration of the classification standards of wearable devices, focusing mainly on software-related device characteristics issues. The study concludes that even though wearable devices have gained momentum in recent years, they are still at an immature stage of development. The authors claim, "Hardware materials and battery life still has not had a breakthrough: limited screen space makes the product design difficult, and application software is still in an initial stage" (p.597).

One observation from these studies is that wearable devices are available in various form factors (size, shape, style, etc.) and that these various form factors and the environments the wearables are used within can affect usability, which in turn impacts on the user acceptance and engagement. Another observation from prior research suggests that previous studies have examined wearable devices and identified usability issues such as screen size, battery life, connection, software and are scattered across the literature, and there are relatively no studies that focus on review and analyze usability and its relation to which types of wearable devices and thus, a need to fill this research gap. Therefore, this chapter aims to fill this research gap by presenting an in-depth, formal, and inclusive review. To present a holistic overview of studies on usability issues related to wearable devices, the present study builds a categorization scheme to identify various types of usability issues and how they have been identified in previous studies.

3   Methods

Based on the guidelines provided by Kitchenham and Charters [26], Engström and Runeson [27], and Petersen et al. [18], an SLR approach was adopted and applied for the current study; these guidelines describe a streamlined SLR approach that researchers follow to gather the necessary data from a pool of scientific literature and how to evaluate and categorize the data in an unbiased way based on the relevancy of the formulated research problem (Kitchenham and Charters [26]). Steiger et al.

[19] assert that "conducting a systematic literature review is an efficient way to select the best available research and facilitates research approaches by identifying current existing research gaps and study limitations" (p.21).

The main advantage of an SLR approach is that it provides information about the effects of a phenomenon across a wide range of settings and empirical methods with the possibility to combine data using meta-analytic techniques [28]. The adopted process consists of the following phases:

- *Define* the research questions (RQs), based on research goals and objectives.
- *Develop* a review protocol that specifies the search, selection, data extraction, and synthesis strategies [29].
- *Conduct* a scientific literature search to identify the primary literature by using generated search strings on electronic databases that consists of articles from conference proceedings, and journal publications. Search string generation sometimes requires an iterative approach before suitable search terms and values can be found.
- *Screen* the preliminary set of identified literature by utilizing inclusion and exclusion criteria (i.e., find the papers that fulfill the objectives given by the research questions, etc.).
- *Categorize* the selected literature based on the set of keywords which is crucial in identifying relevant primary studies [1].
- *Present* the results in a visual form (i.e., in graphs, tables, or other informative graphical representations).

Petersen et al. [18] recommend that researchers doing SLRs should use alternative ways to present and visualize their results. Following the advice and based on extracted data from the selected articles, the current SLR presents the results in graphs, tables, and figures.

3.1   Research Questions

Tosi and Morasca [29] state that "defining research questions is an essential part of the SLR, as they drive the entire review methodology" (p.19). The current SLR identifies usability-related issues and user interface-related issues in wearable devices, investigating how these issues were discovered. Petticrew and Roberts [30] and Kitchenham et al. [31] both suggest using the population, intervention/issue, comparison, and outcome (PICO) framework to formulate the SLR research question. The PICO framework defines research questions by providing the criteria for defining keywords, structuring the final search string, and formulating the inclusion and exclusion criteria. The overall principles of PICO are applicable to any search strategy; however, some PICO elements can be discarded depending on the nature of the research. In the current study, the aims do not include comparing issues related to wearable devices; instead, the focus is on discovering the pertinent issues. Because a comparison is beyond the scope of the current work, this element was thus omitted. Hence, the following four research questions (RQs) were formulated:

RQ1: To date, what categories of usability issues related to wearable devices have been discussed in the past, and which issues relating to wearables still persist and need further investigation?

Rationale: Defines the basis of the SLR, allows us to identify, evaluate, and categorize the range of usability issues and get an overview of the usability issues through a categorization framework (i.e., the issues that have been presented and discussed, along with their implications). The results that answer RQ1 will enable the researchers, practitioners, and application developers to understand and obtain a more holistic overview on which issues currently exist, what caused those issues to appear in the first place, and which issues are associated with which type of wearable device categories. The sub-question of RQ1 provides detailed information on the challenges that still remain and the improvements required to alleviate them, serving as a basis for future research directions. Previous studies indicate that usability as factor that influences abandonment of the devices of the wearable devices; this paper identifies and presents a categorization framework that allows researchers, practitioners, and application developers to understand and obtain a more holistic overview on which usability issues are associated with which type of wearable device categories that act as the barriers to user adoption, facilitating the adoption of wearable device.

RQ2: How have usability evaluation methods (UEMs) been applied to wearable device evaluation and in which device categories?

Rationale: Identifies the range of the most commonly used UEMs and their subsets to obtain an overview on which UEMs have been employed to evaluate the categories of wearable devices. The result obtained from RQ2 will enable researchers, practitioners, and application developers to understand and make decisions while selecting the UEM for a particular type of device evaluation.

3.2 Search Design and Process

The primary studies used in the current paper were identified using search strings on scientific digital databases. In addition, a manual search was done through relevant conference proceedings and journal publications, which is explained in detail below. The automated search process was conducted on the following digital databases: "IEEE Digital Library," "ACM Digital Library," "Springer Link," "Science Direct," and "others: Google Scholar," These databases were selected because they are the preeminent sources of published research in the engineering field. The aim was to find as many notable publications that discuss usability issues related to wearable

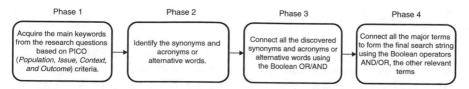

Fig. 1 Search string formulation process

devices as possible. Figure 1 shows the four phases of search string formulation process presented by [1].

In Phase 1, the main keywords relating to the research questions (see Sect. 3.1) were acquired using Population, Intervention, Comparison, and Outcome (PICO) criteria[2] (i.e., "wearable device" and ("usability issue"). Kitchenham et al. [31] recommend the use of keywords from the comparison and outcome criteria when formulating the search string; this was not carried out in the current work because it is only a common procedure in the field of medical science. Kitchenham et al. [31] and Petersen et al. [18] also note that using keywords from the comparison and outcome criteria is not always applicable. In the present case, the use of a comparison was discarded when the research questions were formulated, and the outcome was not taken into account because the current study does not aim to measure effects. In Phase 2, the identification of synonyms and acronyms or alternative words took place. One of the constraints when formulating a search string is that the resulting set should have the maximum possible coverage but should remain at a manageable size. Therefore, several synonyms ("wearable devices," "wearable computing," and "wearable technology$_*$") were used. In Phase 3, Boolean "OR" was applied to merge all the discovered synonyms and acronyms or alternative words [1]. Lastly, in Phase 4, Boolean operator "AND" was applied to connect all the keywords and to formulate the final search string for relevant articles published after the year 2000 as ("wearable$_*$" or "wearable device$_*$" or "wearable computing" or "wearable technology$_*$") AND (""usability issue$_*$" or "usability") AND ("publication year >2000").

In January 2018, an initial search was conducted utilizing the formulated search string and the search utility of the selected digital databases. The final set of searches was performed in February 2018. Additional search was also performed using online web search engine "Google Scholar" to find if any further relevant articles exist and "cross-check the final sets of retrieved papers to determine the relevance of each paper" [2].

---

[2]PICO Criteria: http://learntech.physiol.ox.ac.uk/cochrane_tutorial/cochlibd0e84.php

3.3 Article Selection Process

The article selection process in the current study is defined as a process of extracting the relevant publications with respect to the objective of the SLR based on inclusion criteria (IC) and exclusion criteria (EC). Hence, in this context, the subsequent set of IC and EC were formulated and applied to select the relevant publications:

- *IC1:* Publication is dated between 1/1/2000 and 02/2018.
- *IC2:* It comprises answers to at least one of the presented research questions, which was determined by reading the title and abstracts.
- *IC3:* Publication is written in English.
- *IC4:* If various similar papers are outlined by the same author, only the most current publication is used.
- *EC1:* Publication lies outside the wearable devices domain.
- *EC2:* The publication does not cover the usability-related topic within the domains of wearable devices.
- *EC3:* Technical documentation or reports.

Based on the above ICs and ECs, the article selection process was conducted in four individual phases, as shown in Fig. 2. During Phase 1, an automated search was performed using search strings to identify potential studies. This preliminary search yielded 3271 papers.

In Phase 2, the articles (title, keywords, and abstract) obtained in Phase 1 were reviewed, and the ICs and ECs were applied to select the articles for the next phase of the process. As a result, 350 articles were selected, and 3271 articles were excluded. In Phase 3, a review of the full text of the selected original articles from the previous phase was conducted to determine the articles' relevance and whether the articles should be included for further analysis. Finally, 84 articles were considered suitable, while 266 were excluded because they were not relevant to the RQs, had too little in the way of content, or were not in English (i.e., the abstract, keywords, and title were in English, but the body of the article was in another language). Thus, 84 articles were identified as relevant primary studies for data extraction. Of the 84 studies reviewed, 34 were published in the ACM digital library and rest on the other electronic databases (i.e., Springer, Science Direct, IEEE, BioMed Central, Hindawi, Taylor and Francis, Journal of Medical Internet Research, and Journal of Computer-Mediated Communication), respectively.

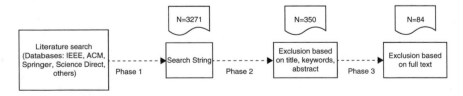

Fig. 2 Article selection process for choosing relevant primary studies

3.4   Data Extraction and Synthesis

Because the results presented in the current paper are based on the qualitative assessment of the previous literature, the process of data extraction and synthesis is described below. According to Welsh [32], in a qualitative data analysis, to avoid human errors and when organizing the data, "it is important that researchers do not reify either electronic or manual methods and instead combine the best features of each" (p.5). Following this recommendation, both computer-based and manual analysis techniques were applied. Furthermore, the approach applied for the data extraction and synthesis process consisted of six phases.

In the first phase, all the relevant articles were exported to the NVivo data analysis tool (version 11) [33] for data analysis from the Mendeley reference management tool [34]. NVivo data analysis tool was applied because it allows for sophisticated data coding and helps map out diagrammatically how the themes relate to each other [32, 35]. After the final set of relevant articles were transferred to NVivo, the initial nodes were created on NVivo based on the main themes: usability issues; usability evaluation method; target group; wearable device categories; wearing position; geographical locations; application domain; and age group. Further nodes were created under the usability evaluation method based on the taxonomy of Ivory and Hearst [36], that is, the method class, method type, automation type, and effort level.

In the second phase, each relevant article was read, and important sections of the text were coded. During the coding process, either phrases, paragraphs, or single words were highlighted and added with links to the initial nodes (i.e., themes) from Phase 1. For example, text from one study that was coded and added within the "usability issues" node could be as follows: "all of them experienced automatic loss of synchronization, making it difficult or impossible to update data or resulting in an incorrect report" [37] (p.8).

To improve accuracy, printed copies of articles were read, and themes were highlighted. Coded data from the NVivo and the highlighted data from the printed copied were compared to see if patterns remained the same on the computer-based and manual method. Some data were missing when analyzed with NVivo. Those data that were missing were added to NVivo. Following this, a node list was generated for debriefing to other researchers. According to Impellizzeri and Bizzini [38], "Data extraction must be accurate and unbiased and therefore, to reduce possible errors, it should be performed by at least two reviewers" (p.499). Based on this recommendation, in the third phase, the initial data sets were reviewed by two members of the research team to confirm that the intended meaning was accurate and appropriate for further analysis. Furthermore, there were no disagreements between the initial datasets.

Because the main goal of the current study is to identify, evaluate, and categorize (i) the usability issues related to wearable devices and (ii) the types of usability evaluation methods that have been discussed in the literature, after the final agreement, in the fourth phase, data related within the node to usability issues and the usability evaluation method were further coded. Nowell et al. [39] note, "Sections of text can be coded in as many different themes as they fit, being uncoded, coded once, or coded as many times as deemed relevant by the researcher" (p.6). For example, for text from one study, "preliminary graphical icons was cumbersome because the icons often ... represent" (p.1125) [40] were coded under theme iconography because it described the usability issues related to the icons.

For the manual approach, each text was coded on a printed list of usability issues that were identified in Phase 2, and a theme was given to each identified usability issue. The final sets of data were exported to excel. Data on excel and Nvivo were compared, showing that some themes on Nvivo were missing during the coding of the primary studies. To obtain the agreement among raters and assess intra-rater reliability, we applied Cohen's kappa, in which two raters separately rate the data. The final percent agreement was 0.976, which we can interpret as almost perfect agreement based on Cohen's suggestion that "values ≤ 0 as indicating no agreement and $0.01 - 0.20$ as none to slight, $0.21 - 0.40$ as fair, $0.41 - 0.60$ as moderate, $0.61 - 0.80$ as substantial, and $0.81 - 1.00$ as almost perfect agreement"(p.6) [41]. A total of 14 data fields were created, which included the following data for each primary source included in the study: study ID (S1, S2, ...), title of the paper, citation, year of publication(s), research focus, type of publication, name of the database where the publication was retrieved from. The data that were relevant to what was obtained through the coding process were exported in excel for further analysis and include the following: usability issues (if applicable), wearable device categories (if applicable), wearing position (if applicable), usability evaluation method (UEM) (if applicable), geographical locations (if applicable), application domain (if applicable), and age group. Extracted data were recorded into data fields and are described in more detail online (https://doi.org/10.5281/zenodo.1476457).

4 Results

This section presents the results that were consolidated from the final set of 84 articles (see Appendix A) based on the RQs formulated in Sect. 3.1. The results are presented in the form of graphs, tables with analysis based on the recommendation by Petersen et al. [18]. As shown in Fig. 3, out of 84 articles, 59 were from conferences (70.23%), and the rest were from journals (29.76%). One reason for this may be because the wearable topic has picked up momentum recently, and conferences have a shorter time to publish when compared with an article. However, the increase in the number of publications shows that the field is becoming more important and that people are paying attention to these issues, which is in line with [42], where they claim, "The significant number of papers in conferences and journals is an indicator that the concept has started to get consolidated" (p.51).

Additionally, the study led to the identification of 19 types of wearable devices utilized in the research articles. The identified devices and how they are distributed are shown in Fig. 4. Most of the studies were carried out utilizing smartwatches,

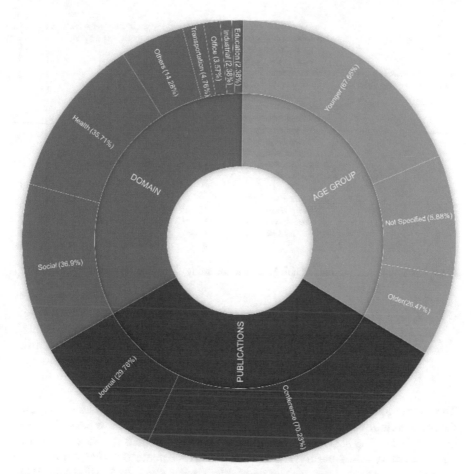

Fig. 3 Descriptive statistics of the publications, age group, and domains of the selected papers

activity tracker/monitor, and wristbands, which are followed by head-mounted displays (HMD) with a binocular configuration (worn over both eyes), either opaque or transparent, for example, virtual reality and smart glasses with AR, and head-mounted displays (HMD) with a monocular configuration (worn over one eye) that is transparent, for example, smart glasses. We speculate that using both commercially off-the-shelf and prototype devices in the selected studies is the reason for having numerous types of wearable device categories.

Additionally, as displayed in Appendix, these identified devices are worn on the wrist (44/84), head (20/84), chest (4/84), finger (3/84), knee (1/84), and the remaining on other parts of the body, such as arms, neck, waist, or feet. The current trend of wearables is mostly wrist-worn and head-worn; however, other body-worn devices are gaining momentum.

In the following section, we discuss how each research question was answered.

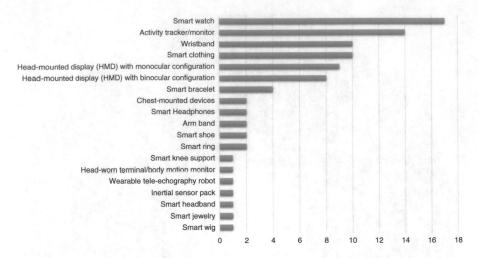

Fig. 4 Categories of wearable technology used in the study

4.1 RQ1: To Date, What Categories of Usability Issues Related to Wearable Devices Have Been Discussed in the Past, and Which Issues Relating to Wearables Still Persist and Need Further Investigation?

The overall aim of this research question was to identify, evaluate, and categorize the set of usability issues related to wearable technologies that have been discussed prior to 2018. The analysis of the primary studies prompted the identification of 20 different types of usability issues related to wearable devices. An example of each in relation to the type of wearable device category is shown in Table 1. As shown in Table 1, most of the identified usability issues are related to smartwatches (15/20), wristbands (12/20), activity trackers/monitors (18/20), head-mounted displays (HMD) with a binocular (worn over both eyes), either opaque or transparent (15/20), and head-mounted displays (HMD) with a monocular (worn over one eye eyes), transparent (14/20). Therefore, we believe the issues related to the devices need immediate attention because these are the most available devices on the market. Similarly, Table 1 also shows that out of the 20 usability issues, screen size, aesthetics (physical design, material, and color), interaction techniques (auditory, visual, gesture, and haptic feedback), wearing position, and motion artifacts were the most reported.

Based on [16, 43], the identified issues in Table 1 were further condensed into device characteristics (see Sect. 4.1.1) and the deployment of wearable devices on the body and external devices (see Sect. 4.1.2).

Table 1 Identified issues in relation to the type of wearable device category

Wearable device categories	Types of associated usability issues																			
	Screen size	Screen display	Lack of screen	Color contrast	Interaction techniques s (auditory, visual, gesture, and haptic feedback)	Button location	Device context	Navigation	Iconography	Elements (fonts/color)	Interaction with the application	Battery	Weight	Memory size	Aesthetics (physical design, material, color)	Wearing position	Motion artifacts	Data accuracy	Device connectivity	Applications on external device
Smartwatch	×	×			×	×	×	×	×		×	×		×	×	×	×	×	×	
Smart wig															×	×				
Smart clothing					×		×					×	×		×	×				×
Smart ring															×	×				
Smart jewelry															×	×				
Smart shoe																×				
Smart bracelet	×				×		×	×	×	×	×				×	×				×
Wristband				×	×	×	×	×			×				×	×	×	×	×	
Arm band												×			×	×				
Smart headband															×		×			

(continued)

Table 1 (continued)

Wearable device categories	Types of associated usability issues																			
	Screen size	Screen display	Lack of screen	Color contrast	Interaction techniques s (auditory, visual, gesture, and haptic feedback)	Button location	Device context	Navigation	Iconography	Elements (fonts/color)	Interaction with the application	Battery	Weight	Memory size	Aesthetics (physical design, material, color)	Wearing position	Motion artifacts	Data accuracy	Device connectivity	Applications on external device
Smart headphones	X				X									X						
Inertial sensor pack		X	X		X												X			
Activity tracker/monitor	X	X	X		X	X	X	X	X	X	X	X	X		X	X	X	X	X	X
Wearable tele-echography robot																x				

Head-worn terminal/body motion monitor	X		X			X				
Smart knee support						X		X		X
Chest-mounted devices		X	X	X						X
Head-mounted display (HMD) with binocular configuration (worn over both eye) opaque or transparent, for example, virtual reality, smart glasses with AR	X	X	X	X	X	X	X	X	X	X
Head-mounted display (HMD) with monocular configuration (worn over one eye) transparent, for example, smart glass	X	X	X	X	X	X	X	X	X	X

Device Characteristics

Ally and Gardiner [43] specify that smart mobile computing device characteristics can be classified mainly by two components: physical and user interface aspects. Specifically, the physical component concerns product aesthetics that relate to the external look and feel and internal components, such as sensors, processor, memory, power supply, and transceiver [43, 44]. Lee et al. [4] define the user interface as the way in which users interact with devices while managing their interactions with other machines and the people who are connected to the device.

The categorization resulted in 15 issues related to the device characteristics (out of which 20 were related to the user interface and four to the device's physical aspects). We further clustered the user interface issues based on the user interface fundamentals explained by Dennis et al. [45], who state that "the user interface includes three fundamental parts: the output mechanism (the way in which the system provides the user with information), the input mechanism (the way in which the system captures the information), and the navigation mechanism (the way in which the user gives instructions to the system)" (p.314). The input and navigation mechanisms are combined in the current work because both mechanisms relate to receiving the instructions and system capturing from the user. Nine issues regarding the input and navigation mechanisms were found, and five issues related to the output mechanism were discovered. Moreover, two issues were found to be associated with both the Output mechanism, input and navigation mechanism aspects of the device. A summary of mapped usability issues associated with the device characteristics is shown in Table 2.

Each of the usability issues relating to device characteristics (both user interface and physical) are discussed below to clarify how they impacted the use of wearables among individuals.

Screen size: Considering that most, if not all, other computing devices have screens, wearables are an oddity in that many of them do not have screens. Devices with screens play a vital role in providing better user-device interaction and increase user engagement while delivering content such as quantified-self data, online manuals, and training tools with augmenting devices [24, 46]. Dhawale and Wellington [24] find that screen size was significantly important for participants because "it makes the user interaction with the device easier, smoother and more engaging" (p.41). However, to deliver the wearability, portability, and fashionable characteristics required [47], wearables are designed with very limited display size and shape. For example, head-mounted devices are designed with limited display and shape, and the visual field is only a small central region (23 degrees) [48]. Having such a limited display size and shape restricts the input, output, and navigation capabilities of the devices. Wichrowski et al. [49] use head-mounted display (HMD) with monocular (worn over one eye) configuration for example, Google Glass as a wearable device and find that "screen size is too small to convey a fairly substantial amount of information" (p.4). Similarly, Kim [50] find that screen size and shape affects the information quality and inhibits the content-

Table 2 Summary of the mapped usability and user interface issues associated with wearable devices related to device characteristics (table representations)

Usability issues categories	Usability issues subcategories	Fundamental	Associated usability issues
Device characteristics	User interface	Output mechanism	Screen size
			Screen display
			Lack of screen
			Color contrast
			Interaction techniques (visual, auditor, and haptic feedback)
		Input and navigation mechanism	Screen size
			Interaction techniques (gesture, auditory)
			Button location
			Device vontext (text, time, and visualization)
			Navigation
			Iconography
			Elements (text/button)
			Interaction with the application
			Color contrast
	Physical	External look and feel	Aesthetics (physical design, material, color)
			Weight
		Internal component	Battery
			Memory size

relevant thoughts among users. For example, Kim [50] report, "The large screens inhibited participants from generating thoughts about the specific contents of the given information, such that the messages presented on the large screens elicited fewer content-relevant thoughts than the messages presented on the small screens" (p.131). Moreover, having a small screen size not only limits devices to delivering information, but it also reduces the usability while interacting with the devices and the individuals' intended goals while performing certain tasks, such as reading the messages, navigating within the application, and typing messages. For example, Pulli et al. [51] note the issue of display size: "The display size was too small for easy reading" (p.1125). This can increase the number of user errors, affect efficiency, and alter the individual's decision to continue using the device for a longer period of time [24, 52].

Screen display: The screen display is one of the more influential factors that causes usability issues among individual's interactions with the device because of the display's size, position, or shape [49, 53], technology, such as small prismatic crystal [49], or configuration, such as monocular (i.e., worn over one eye) and transparent or binocular (i.e., worn over both eyes) and transparent [54–56]. For example, Delabrida et al. [54] report that the lenses of head-mounted wearables with an opaque binocular display configuration increased the smartphone screen resolution, leading to higher image resolution and a usability issue among participants. Similarly, Laramee and Ware [56] find that wearable devices with a transparent monocular display configuration negatively impacted the task performance, such as reading and viewing against anything other than uniform background. In addition, McGill et al. [57] report that participants found the opaque view quite disruptive while using HMD devices. Kaewkannate and Kim [37] find that for participants, it was difficult to see the text in sunlight using the current wearable displays.

Lack of screen: To explore new possibilities when it comes to making devices lighter, many device manufacturers have physically discarded the screen; only sensors are used to track the users' daily activities, such as sleep and physical activity, and the data are presented through external devices, such as a smartphone or computer. This state-of-the-art technique has the advantage of utilizing a non-visual user interface (UI) in terms of (i) wearability, for example, being light weight; and (ii) in reducing issues with remembering individuals, for example, charging devices [58]; however, the study identifies that having a non-visual interface brings additional usability challenges. For example, Kaewkannate and Kim [37] find that there are issues related to interaction with the devices (i.e., the device does not respond while tapping its surface); needing to use the apps on external devices to view the information; and data inaccuracy because of the synchronization between the external devices and the wearable.

Color contrast: Color contrast assists individuals in viewing and interacting with content; it consists of elements (i.e., color, text, and graphics) in a dark and bright-light environment. However, the current review indicates that (i) poor color contrast between the background of the user interface and the color of the text affect readability [59] and (ii) visual cues with a higher color contrast have less reaction time among participants [60]. For instance, in the study by Holzinger et al. [59], a participant commented: "Dark grey text? On a light grey background, it is very difficult to read" (p.4). Similarly, Costanza et al. [60] find that when visual cues with a bright color contrast were delivered on the eyewear display of the user, the reaction time was quicker than when there were visual clues with a dim contrast. Wichrowski et al. [49] also state, "All personal graphic elements must meet the requirements of usability. For example: too detailed graphic elements or too many colors can interfere with readability and may not be visible on Google Glass. Therefore, it is necessary to create graphics from a limited number of elements and colors" (p.3).

Interaction technique: Jacob [61] state, "An interaction technique is a way of using a physical input/output device to perform a generic task in a human–computer dialogue and represents an abstraction of some common class of interactive tasks, for example, choosing one of several objects shown on a display screen. Research in this area studies the primitive elements of human–computer dialogues, which apply across a wide variety of individual applications" (p.1). Many of the review papers reveal that wearable devices, from smartwatches to the HMDs, provide many different ways of interaction through several in- and output modalities, including auditory [62], visual [63], haptic feedback (i.e., tactile, kinesthetic) [64], and gesture (i.e., touch, head) [65–68], to enhance the user experience.

Although these interaction techniques have bought new opportunities when it comes to improving the user experience within the interaction domain, the current study shows that it also opens up usability challenges. For example, Kaewkannate and Kim [37] find kinesthetic feedback that relates to force feedback sensed from muscles, joints, and nerves when the user receives output from the device, such as vibrations caused irritation among users during evaluation. One of the causes was because of excessive vibration before the device went into sleep mode. In addition, both Lazar et al. [6] and Zhang and Rau [69] report concerns from participants regarding receiving active feedback, that is, notifications from devices while performing a task. For example, participants commented, "I don't want to receive messages when I am doing the exercise" [69].

Regarding the touch gesture, studies report the reasons why users encountered problems while interacting with the device's surface. For instance, the studies note that maintaining touch [70] and touch sensitivity on the device [71, 72], which require frequent tapping on the device surface [68], were some challenges users encountered with tapping. In addition, Pulli et al. [51] analyze tap detection, finding that when users selected a too low detection threshold on the device, they suffered slight pain in their finger. Other studies show that there may be additional challenges related to tapping. For instance, Wichrowski et al. [49] find some participants reported swipe gestures caused challenges because of the placement of the device, for example, "for many students it was not natural to use swipe gestures close to the head" (p.9).

In line with the above challenges, studies indicate that poor speech recognition [49, 62, 63] by the device and needing silence for speech input [62] were the main challenges users discovered while conducting auditory interactions. For example, Neto et al. [62] report that users experienced difficulties while interacting with the device when starting the gear face recognition (GFR) system through speech synthesis. Similarly, Lawo et al. [63] find that using voice as the interaction technique was difficult for participants with an accent. Wichrowski et al. [49] also observe that Google Glass sometimes incorrectly interpreted voice commands.

Button location: The button acts as the meditator between the user and wearable devices, in this case being the item responsible for triggering actions. A user can touch the button either to give instructions or navigate the content within the device. The location of the buttons on both the hardware or on the screen of the wearable devices can affect users' actions and may lead to serious mistakes. Rasche et al. [73] point out, "The activity tracker just had one button to interact with. Participants reported this interaction design to be difficult and annoying. They had problems feeling the button under the silicon tracker display wristband. The navigation of the activity worked by pushing the button. For example, it was necessary to press the button twice to get the actual time, which was reported to be annoying" (p.1414). Ye et al. [74] also "observed participants feeling along the far side of the wrist to find the top edge of the "Select" button or along the near side to find the bottom edge of the 'Home' button" (p.7). Holzinger et al. [59] assess the behavioral intention and user acceptance of a wrist device. They find that the manual alarm button of the device they study "tended to stick, requiring more pressure to activate than was compatible with an elderly person in need of help" (p.4).

Device context (text, time, and visualization): The literature review indicates that issues surrounding the context of the device were influenced by the text, time, graphics, and visualizations during the interactions with wearables. For example, Altenhoff et al. [75] observe that participants had issues using devices when they realized that there was no option to input decimal amounts (e.g., 10.5 ounces of water). Rasche et al. [73] find that participants would substitute their wrist watch if the trackers permanently displayed the time of day. Ananthanarayan et al. [76] observe that participants found circular LED displays were harder to count and calculate. Device with the Knee-shaped visualizations received a mixed response. The non-linear display was found to be complicated and aesthetically unpleasing among participants.

Navigation: The literature review shows that navigation allows users to easily access the information using (i) touch technologies, such as analog resistive and capacitive, through swipe and tap to navigate within the user interface; (ii) a control button on the side to navigate within the user interface; and (iii) capacitive touchpads with or without light-emitting diode (LED) patterns in conjunction with external devices [37, 75, 77, 78]. However, at the same time, small screen sizes, gesture, buttons, or the lack of a screen, and the limits of user usage behavior [79] (i.e., keep their eyes locked on the device while conducting interactions) impact the efficiency and usability, which includes swiping, difficulties locating settings, menus or icons, and unintentional interruptions during interactions. For example, Thorpe et al. [80] find that participants had difficulties "to know when to stop swiping through the menu on wristwatch" (p.300). Altenhoff et al. [75] discover that several of the Jawbone Up participants had trouble locating and understanding the alarm. Furthermore, one participant was confused about the "smart sleep" setting: " .... but is that before or after?" (p.243). Wulf et al. [81] observe that "initiating the speech interaction by pressing the physical button sometimes led to the problem that the participants forgot to push the button and started speaking without the system listening to them ... nor could receive the user's command ... Moreover, the interaction was sometimes interrupted unintentionally by pressing the activation button again so that the previous conversation and dialogue was erased and the system got restarted" (p.204). In addition, Kaewkannate and Kim [37] show that it is difficult to navigate simultaneously on an app installed on external devices and one installed on wearable devices.

Iconography: The results from the previous studies [51, 75, 80] indicate that an icon allows users to (i) launch an application on devices and (ii) navigate within a user interface to locate functionalities. However, at the same time, the studies on this point out that unintuitive application icons or a poor presentation of the icons on without a text label on an application has a negative impact on the usability. For example, Thorpe et al. [80] show that it was difficult for participants to find a functionality based on the icon's name application on a device user interface. Similarly, Pulli et al. [51] find graphical icons without a label give users more problems when it comes to recognizing their clear association with the function behind the icons in a first-time interaction with the application, for example, a "mobile phone" icon was confused with a "door icon" [51] (p.1125).

Elements (font/button): The results show that the form factor of the devices, such as smaller font size, made it harder for participants to read the devices' screens. For example, Holzinger et al. [59] report that participants wanted to adjust the size of the text when they were not wearing their glasses. Wichrowski et al. [49] also show the suggestions they received about using larger fonts for better readability.

Interaction with the application: According to Carter [82], the usability challenges associated with interactions with the application developed because of too many steps that the user had to go through, which deterred the user from completing the task. Other than issues caused by too many steps, Altenhoff et al. [75] note issues that were influenced by the data provided by the wearable device, which affected the users' first experiences with the application; they further explain that users' first impressions of the application may have lasting effects on user engagement.

Battery life: All wearable devices require higher processing power to accumulate and process data through multiple sensors. In addition, the devices consist of smaller battery sizes that are bottlenecked by the wearable device's shape, weight, and size [23]. In addition, the devices are either integrated with a display or require external devices with Bluetooth connectivity to show processed data. This results in higher battery consumption and limited battery lifetime. Ahanathapillai et al. [83] state that the "hardware used... is limited in battery life and offers between 5 and 8 hours when used continually... [this] is recognized as a significant limitation of the hardware" (p.28). Similarly, Sultan [84] points out, "Battery life need to be longer than what is currently available" (p.525). Yang et al. [85] find that an issue with battery life is interrupting data collection, and as a result, devices showed lower quantified-self data, for example, number of steps. Shih et al. [58] and Koskimäki et al. [86] note that issues related to battery life required the user to partake in an additional behavior, such as remembering to charge the device and putting the device on after recharging. For example, Shih et al. [58] state how one participant stated, "It was quite annoying for me remember to wear the device and also to charge it every day" (p.6). Similarly, Thorpe et al. [80] find a higher consumption of battery life impacted older users with dementia because of their challenges in locating the charging port and inserting the cable. In another study, Albrecht et al. [87] find the battery life of smart glasses drained quicker than other devices, such as camera, which was used in parallel during the experiment. They point to the concern of human battery interactions (HBI) [88]. For example, "In medical settings, a cable running down from Glass to an external battery pack may raise concerns with respect to hygiene as well as add potential for Glass to be pulled off the user's nose if something (e.g., a fastener on the physician's surgical gown) inadvertently pulled on this cable" (p.11).

Weight: Although the main aim of wearable devices is to provide users with wearability and portability, the current literature review identifies "weight" as one of the influential usability issues for individuals. Furthermore, the results indicate that the impact of the weight of the devices increased the feeling of attachment among individuals which relates to discomfort [89]. It has also been shown that weight influences device usage. For example, participants in the study of Dhawale and Wellington [24] stated, "Smartwatches to be distracting and bit heavy after a long day of using it" (p.42). Further, Spagnolli et al. [90] assess user acceptance of wearable symbiotic devices and emphasize " . . . lighter and more discrete wearable devices are better appreciated compared to bulkier and more noticeable ones" (p.96) for individuals.

Memory size: The amount of memory required in wearable devices increases when the device runs a more memory-intense operating system, more applications, and performs heavy tasks (i.e., fetching and parsing data from a cloud or external devices and background apps). Less memory means limitations on the type of applications and tasks that can be performed. According to Oakley et al. [91], when designing applications for low-end devices with slow central processing units (CPU), manufacturers and content providers should consider utilizing comparative textual feedback because these devices may not be able to smoothly process image or video information. Delabrida et al. [54] state that memory events represent the main bottleneck in their experiments.

Aesthetics: Wearable devices that are worn outside the body are considered high-tech devices and fashion accessories [47]. The current literature review identifies, irrespective of age, that this dual consideration impacted usability, which is mainly influenced by aesthetic elements, that is, physical design, color, and materials of the wearable devices during human–wearable interaction. For example, both Rodríguez et al. [71] and Ananthanarayan et al. [76] find that the physical design of the device caused difficulties for participants when attaching the devices to their bodies. Ju and Spasojevic [92] also uncover that the design played an important role for the acceptance of smart jewelry. Similarly, Shih et al. [58] show participants felt the device was cumbersome and intrusive when worn during their daily activities. Ye et al. [74] show how participants also suggested changes with the design, for example, being easily customizable, after using the prototype wearable device. The wearable device's size ultimately determined the degree of daily use. Goto et al. [93] show how participants stated that the size of the watch was one of the constraints that affected its usability; the problem was that the size of the device was inconvenient for users to wear in their daily activities. In agreement with Goto et al. [93], Kondo et al. [94] find that the device size was too large for the participants to use. In other study, Nirjon et al. [95] note how the participants commented that the size of the used wearable ring was larger than a typical ring, making it uncomfortable for all-day wear.

In addition, Abbate et al. [96] show that the shape of the wearable device impacted the usability among participants, here mainly because of the bulkiness and shape of the devices during sleep, which initially made it almost impossible to carry out sleep tests. Abbate et al. [96] state that when they repeatedly modified the device and moved the battery/transmitter, elderly users enjoyed wearing the wearable caps at night. Further, Abbate et al. [96] also suggest that ergonomic and aesthetic modifications would be necessary for improving the level of usability and acceptability, especially in an elderly user population: "... the elderly are attached to a specific aesthetic dress code, characteristic of their likes/dislikes ... [they] prefer simple, loose, and comfortable dress, and therefore, the focus should be on a retro style" (p.6). Abbate et al. [96] also find that providing devices in the participants' favorite colors made the wearable devices more acceptable.

Moreover, the current literature review shows that the material and color used on the devices also impacted usability. For example, the wearable devices utilized by [86] for early detection of migraine attacks caused skin irritation among the participants. Similarly, Rodríguez et al. [71] discovers that materials and color applied on the device also impacted usability. In their study, participants suggested the use of different colors or plastic material on the devices. Rapp and Cena [4] find that the color of the tracking devices made it difficult for participants to integrate the device into their daily lives. For example, participants pointed out, "A light blue bracelet could fit with a casual dress for going out with friends, but not with a night dress for formal situations, where she liked more unnoticeable colors" (p.142). In addition, Brun et al. [65] state, "due to the chosen material, the utility stretch straps were judged too small not big enough for some heads, or it could stick with long hair" (p.7), which caused the participants discomfort.

Deployment of Wearable Devices and External Devices

Liu et al. [16] classify each technology by its location on the body. According to Liu et al. [16], "Technology can be on the body (such as wearables), inside the body (such as implants), and carried next to the body (smart phones)" (p.2). Employing a similar classification approach, the issues noted when reviewing the papers were clustered into categories related to the deployment of wearable devices on the body and the corresponding categories: external devices that are carried next to the body (e.g., smartphones). The results of the classification summary are shown in Table 3.

Table 3 Summary of the mapped usability and user interface issues associated with wearable devices related to external devices and deployment of wearable devices on the body (Table representations)

Usability issues categories	Usability issues subcategories
External devices and deployment of wearable devices on the body	Wearing position, motion artifacts, data accuracy, device connectivity, applications installed on external devices

Each of the usability issues related to the deployment of wearable devices, including external devices, are discussed below to clarify how they impacted the use rate.

Wearing position: Wearable devices are either worn on the outside of the body in the form of a watch or on the inner part of the body as an implant. Although the goal of the wearable device is to provide better wearability when the device is worn on different parts of the body, Fang and Chang [97] show that the wearing position of wearable devices impacts individuals' interests, anxiety, visibility, and readability. Similarly, Rodríguez et al. [71] report that when device was worn on the waist during experiment, readability was low. In addition, other studies emphasize that performing a task could be impacted by the wearing position. For example, Carter et al. [82] note how it was difficult for participants to interact with an application to type with the watch hand and difficult to lift the arm to view the screen. Additionally, Chen et al. [79] say that although individuals had accurate results from their physical activity while using wearable devices, such as pedometers and accelerometers, when the devices were worn on the waist, they had difficulties during toileting and dressing. Zhang and Rau [69] also report that many participants could not see the content on the watch clearly when jogging because of body vibration and distortion. Moreover, Nirjon et al. [95] state that when using a wearable ring, placing the fingers and palm flat on a surface impacted the user while typing.

In addition, there are usability issues from the wearing position that are caused by a halo-effect between the individual and device characteristics. For example, in the study by Rodríguez et al. [71], participants indicated that where the device was positioned on the waist did not adjust to meet heavier body types: "I am bigger and fat" (p.8). Mizuno and Kume [98] observe a similar pattern while evaluating a glasses-like wearable nasal skin temperature measurement device: "For some test subjects, differences in the distance from the thermopile sensors attached to the glasses and the nose or forehead caused by head or face shape variations prevented appropriate measurements" (p.730). Similarly, Yoo et al. [99] note that devices such as a watch or a wristband, which can be worn on the wrist or waist (e.g., abdominal binder), are more easily accommodated than a head-worn device. For example, they state "the head is hardly suitable for patients because they don't frequently wear even a hat" (p.364).

Motion artifacts: To recognize users' activities while at a state of rest or in motion, wearable devices are embedded with a network of sensors; here, designers make the assumption that the device is worn in the predetermined orientation position relative to the individual's body [100]. However, the predetermined position may gradually change because of an incorrect wearing position or wrong body movements while at a state of rest or in motion. Although motion artifacts do not have a direct impact on the individual, the current study shows that motion artifacts have a usability effect through the quality of data delivered, that is, through inaccurate data. For example, Ahanathapillai et al. [83] measure the parameters from an accelerometer on the wrist as an indicator of wrist movements. The results show that the measurement changed from medium to high with lots of wrist movement, rather than a recording of low activity data. Moreover, Ahanathapillai et al. [83] also report the presence of motion artifacts because of the movement of the fingertip of the subject, which, in this case, may have produced inaccurate heart rate measurements. Similarly, Klingeberg and Schilling [101] also find arm movements caused distortion on pressure signals. In other study, Chen et al. [79] report that the number of steps captured by wrist-worn devices was more than the participants actually walked, stating that this may have been "because that the wearable device used in this study took into account the arm movements. Therefore, it is unclear how much counts come from walking" (p.37).

Data accuracy: Wearable devices capture and provide digital data, such as quantified-self, image, location, audio, and video with the help of embedded sensors and camera. Those data are either used by individuals to monitor their activities or track their health conditions for their well-being or for entertainment purpose. However, the literature review shows that usability is disrupted by a deluge of inaccurate data caused by (i) motion artifacts; (ii) device connectivity; and (iii) physical conditions. For example, in the study of Liang et al. [102], the participants reported wrist-worn devices were shown to be "asleep" while "reading," causing "issue of trust" among users. Altenhoff et al. [75] observe that participants had problems trusting sleep data after the first night a device failed to accurately report the time they fell asleep. In the same study, one participant was surprised when upon first syncing the band and app, the app displayed about 80 steps before she had taken any actual steps, which she then commented on the third day, "I feel like I would just use it when working out to figure out what I'd actually done and for sleep but not walking because it's not accurate" (p.244). One participant in a study by Kaewkannate and Kim [37] responded, "the display to check the tracking status requires a smart-phone. Sometimes, data are inaccurate because of lost syncing to the smartphone" (p.8). Masai et al. [103] report that sensor data were saturated when the sensors were exposed to ambient light, that is, sunlight, which caused a smart eye wearable to deliver incorrect data of the wearer's facial expression.

Device connectivity: Most wearable devices do not include a built-in global system for mobile communications (GSM) or a global positioning system (GPS) module; they often pair with external devices such as smartphones or a computer using Bluetooth or Wi-Fi to exchange data and deliver relevant information. However, the results from the review show that the initial pairing between devices is either difficult, or when paired, the connectivity is unreliable. For example, Wichrowski et al. [49] study Google Glass and had problems pairing the wearable with smartphone devices via a Bluetooth connection. According to Wulf et al. [81], a steady and reliable Internet connection could substantially increase the usability for speech-only interactions for wearable systems; this unreliable connectivity resulted in inaccurate data. Similarly, Kaewkannate and Kim [37] report that the automatic loss of synchronization between wearable and external devices made it difficult to update data or resulted in inaccurate data. This demand for a connection impacted the usability of users. Moreover, Thorpe et al. [80] observe that the Bluetooth connection between the wearable and external device dropped unexpectedly, requiring participants to reset the wearable device. Further, Rasche et al. [73] observe that even though the necessary graphical interface was integrated into the app, the installed application on external devices demanded a Bluetooth connection, which was difficult to handle for most participants.

Application installed on an external device: The literature review shows that because of technological and design challenges, the wearable devices used in the previous study mostly work in parallel with mobile devices. For example, when a certain action is performed on wearable devices, such as fitness tracking, the reaction is displayed on an application installed on a mobile device. However, the current analysis shows that having to use applications installed on external devices imposes higher mental effort and stress for users because of concerns related to interruption, installation, and actions that must be performed. For example, Ananthanarayan et al. [76] find that with smartphones as the external device, participants were concerned about being interrupted by texts and phone calls, losing focus, and having to prop up the phone for a better viewing angle. Rasche et al. [73] observe that participants needed to maintain a higher mental effort to install the app on external devices and get the activity tracker to work. Rapp and Cena [22] show that one participant became stressed about the action she had to perform to view the data, for example, "take out the phone, open the app, and explore the graphs and numbers, which she termed 'a laborious task'" (p.141). In another study made by Dhawale and Wellington [24], participants indicated concern regarding having an application on the external devices and viewing this application while performing a physical activity because of what the participants perceived as risk associated with the device screen getting damaged. In addition to this, Kaewkannate and Kim [37] evaluate wearable devices with an application installed on the external devices, and they emphasize the usability concerns related to interacting with the application that is installed on external devices, for example, difficulty with using the food log and calorie tracking tool in the user interface (UI) of the application.

In summary, following Ally and Gardiner [43] and Liu et al. [16], the categorization resulted in 18 issues related to the device characteristics, including shared issues between UI fundamentals, three issues related to the deployment of wearable devices on the body, and three issues related to external devices, in which data accuracy is added separately to each category. Although all categories were important, analysis shows that issues associated with the deployment on the body, external devices, and physical (i.e. aesthetics) have the most influences on user interaction.

Combining the identified usability issues, it is useful to understand and assess the relationship between the usability issues related to the device characteristics, deployment of the wearables on the body, and the use of external devices; to do this, the constructed categorization framework is given in Fig. 5, which combines a total of 20 usability issues and gives the holistic view of overall usability issues that currently exist in all type of wearable devices. Furthermore, the categorization framework clearly shows that some of the usability issues related to the wearable

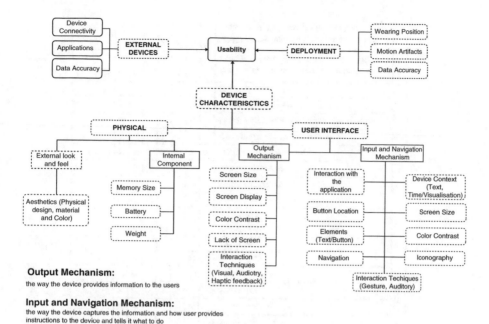

Fig. 5 Usability issues categorization framework based on the reviewed paper

devices share common themes across other categories (Fig. 5). For example, both device connectivity and deployment share the cause of data inaccuracy.

However, the categorization framework does not show which type of issues categories and subcategories is related to what type of wearable device. Therefore, the categorization framework was further reviewed in relation to the wearable device categories presented in Table 1. Table 4 provides the overview of each type of wearable device category, usability issues categories, and subcategories and associated usability issues.

4.2 Q2: How Have Usability Evaluation Methods (UEMs) Been Applied to Wearable Device Evaluation and in which Device Categories?

This section summarizes which evaluation methods have been applied to identify the issues discussed in Sect. 4.1. To answer the first part of RQ2, the methods reported in the primary studies were analyzed and grouped into a taxonomy, as proposed by

Table 4 Summary of the usability issues based on the categorization framework in Fig. 5

Usability issues categories and subcategories			Associated usability issues	Smart watch	Smart wig	Smart clothing	Smart ring	Smart jewelry	Smart shoe	Smart bracelet	Wristband	Arm band	Smart headband	Smart headphones	Inertial sensor pack	Activity tracker/monitor	Wearable tele-echography robot	Head-worn terminal/body motion monitor	Smart knee support	Chest-mounted devices	HMD with binocular (worn over both eyes) opaque or transparent, for example, AR	HMD with monocular (worn over one eye) transparent, for example, smart glass
External devices			Data accuracy	X												X						X
			Applications			X			X							X			X	X		X
			Device connectivity	X							X					X					X	X
Deployment			Data accuracy	X							X					X					X	X
			Motion artifacts			X	X		X	X	X	X	X		X	X		X	X		X	X
			Wearing position	X	X	X	X	X	X	X	X	X	X			X	X				X	X
Device characteristics	Physical	External/Look and feel	Aesthetics (physical design, material, color)	X	X	X			X							X		X			X	X
			Weight			X										X					X	X
	Internal/Component		Battery	X								X									X	
			Memory size	X											X							X

		1	2	3	4	5	6	7	8	9	10	11	12	13
User Interface	Input and navigation mechanism													
	Iconography	X						X			X			X
	Color contrast													X
	Screen size				X			X			X		X	X
	Device context (text, time/visualization)	X		X		X		X	X		X		X	X
	Navigation	X		X		X		X			X		X	
	Elements (text/button)	X				X	X				X			
	Button location	X				X		X			X			X
	Interaction with the application	X				X	X	X			X			X
	Interaction techniques (gesture/auditory)	X			X					X	X			X
Output mechanism	Interaction techniques (visual, auditory, haptic feedback)					X	X	X	X		X	X	X	X
	Color contrast					X	X						X	X
	Lack of screen										X			
	Screen display	X						X			X	X		X
	Screen size	X				X		X			X			X

Ivory and Hearst [36]. Accordingly, the UEMs can be grouped into four dimensions, as follows:

Method class: This comprises the method, such as usability testing and simulation, and is entirely conducted at a high level. This usability evaluation method can be further classified into the following five classes [36]:

- *Testing:* With the intent of finding the usability issues, an evaluator watches user while they are using the evaluated applications/devices.
- *Inspection:* An evaluator creates and utilizes a set of evaluation guidelines or heuristics to assess the possible usability issues related to applications/devices.
- *Inquiry:* The extent to which users share their usability experiences with the evaluators regarding the evaluated applications/devices via methods such as interviews or surveys is examined.
- *Analytical modeling:* This is the degree to which an evaluator predicts the usability issues of the evaluated applications through modeling tools.
- *Simulation:* This is the extent to which an evaluator discovers the usability issues by deploying simulation tools of the applications as if the user is interacting in reality.

Of these five method classes, "testing, inspection, and inquiry are suitable for both formative evaluation" [36] (i.e., the evaluator identifies specific usability problems that are already known before conducting the evaluation) and summative evaluation (i.e., the evaluator obtains general evaluations of usability) purposes, where there are "analytical modeling and simulation" are appropriate for the performance evaluation of users.

Method type: This represents how the usability evaluation (UE) is performed under the method class and with a range of UEMs, such as performance measurement and think-aloud.

Automation type: This represents the use of highly automated techniques for the UE in which a software tool is utilized to simulate the user's action in capturing the data, for example, the software tool automatically records the usability data by logging the user interface usage [36], analyzing it (i.e., the software installed on the devices automatically records usability issues), and critiquing it (i.e., the software installed on the devices analyze the usability issues and suggest improvements).

Effort level: This is the level of human effort required while executing the UEM dimension (method class and method type). The effort level can be (i) minimal effort (MF) (i.e., does not require interface usage); (ii) formal use (F) (i.e., requires completion of specifically selected task); (iii) informal use (IF) (requires the completion of a freely chosen task); or (iv) model development (requires the evaluator to develop the UI model to employ the method) [36].

Table 5, which represents the answers to the second part of RQ2, contains wearable categories, usability evaluation method type (UEMT), usability evaluation method class (UEMC), automation type, and effort level. As shown in Table 5, a

Table 5 Studies reporting the use of each evaluation method

Method class	Method	Automation level	Effort level	Smartwatch	Smart wig	Smart clothing	Smart ring	Smart jewelry	Smart shoe	Smart bracelet	Wristband	Arm band	Smart headband	Smart headphones	Inertial sensor pack	Activity tracker/monitor	Wearable tele-echography robot	Smart knee support	Head-worn terminal/body motion monitor	Chest-mounted devices	HMD with binocular (worn over both eyes) opaque or transparent.	HMD with monocular (worn over one eye) transparent.
Inquiry	Interview	N	F, IF	x	x	x				x	x					x					x	x
	Questionnaire	N	IF	x		x	x	x	x	x	x							x	x		x	x
	Diary	N	IF					x	x		x					x						
	Survey (pre-post)	N	F	x		x					x					x					x	x
	Observation	N	F	x				x		x			x								x	
	Self-reporting logs	C	F								x						x				x	
	User feedback	N	IF			x					x											
	Focus group					x			x							x						x
Testing	Think aloud protocol	N	IF		x											x		x				
	Log file analysis	N, C	M, F	x			x				x				x	x	x		x	x	x	
	Performance measurement	N, C	IF, F	x							x			x		x			x	x	x	
	Question-asking protocol	N	IF								x					x						x
Inspection	Feature	N		x																	x	
	Perspective based	N																	x			

review of the 84 studies revealed that 78 studies did include a UE. In the studies, 14 different method types were applied to understand the usability of the 19 types of wearable device categories. Regarding the automation type, three studies apply the application to record the usability data (C), and others do this using an evaluator with either freely chosen task or specific selected task without any level of automation supported (N). The literature review indicates that the studies have adopted effort level (i.e., formal use, and informal use, model use). To gain an overview of how the UEM type was conducted within a method class, the obtained method types were grouped based on five method classes. After grouping, three method classes were identified: inquiry, inspection, and testing. However, the results show that none of the studies applied analytical modeling.

More specifically, we can see from Table 5 that out of 14 obtained evaluation method types, 41 studies adopt multiple evaluation methods to gather multiple sources of data to get a better overview of wearable devices from the users' perspective. In most of the studies, the interview UEM type was applied during usability evaluation sessions. For example, Altenhoff et al. [75] apply multiple evaluation methods, including think-aloud, post-survey, and interviews (i.e., unstructured), to evaluate two different activity trackers and their associated applications. The think-aloud protocol allowed participants to speak and perform the task by, for example, setting up an activity tracker device and its associated application, allowing the evaluator to collect data such as time-on-task and average error. To gather the overall experience of the device usage, and eliminate the issues of reactivity, participant's verbal abilities, and validity [104], researchers further apply additional usability methods, such as a through post-survey and interview.

Additionally, Rasche et al. [73] evaluate the usability of the activity trackers by utilizing DIN ISO 20282-2 and applying the think-aloud protocol. Because the think-aloud protocol itself cannot grasp the mental effort of the participants, the researchers adopt the Rating Scale of Mental Effort (RSME) unidimensional instrument. Moreover, Rasche et al. [73] use interviews and different questionnaires; first, the Post Study System Usability Questionnaire (PSSUQ) is used at different times to understand the participants' attitudes about the product and changes in perceived usability [105]; the MeCue questionnaire is also used "to evaluate the perceived aesthetics of the activity tracker, the stigmatization of using it, the wearing position, and the intention of usage" (p.1412); a technical affinity questionnaire is used to understand if technical affinity changes during the process of getting used to the application by participants. In summary, having different questionnaires allowed the researchers to gather data from different angles and understand the participants' attitudes toward usability, requirements, motivation, mental effort, and technical affinity of activity tracker. Additionally, Fang and Chang [97] use a pre-test questionnaire to finalize the contents of the formal questionnaires.

On the other hand, the current study also identifies that the type of experimental tasks and period of use of devices also influences how users perceive the hedonic and pragmatic values of the evaluated devices or user interfaces [50]. In the reviewed papers, the researchers who adopted longitudinal usability testing with informal use collect more data to understand what effect the adoption of the device or user interface has. For example, Shih et al. [58] gather the logs of the usage data from activity trackers, collecting the usage pattern and issues of remembering, physical design and aesthetics, data management, integration and sharing, and data accuracy. Lazar et al. [6] discuss the advantages of conducting a long-term usability evaluation with a freely chosen task and selected device while conducting the interviews later, here stating, "By allowing participants to choose devices and then interviewing them several months later, we were able to see the ways people integrated devices into their lives or abandoned them and the factors for doing so" (p.644).

Apart from the adoption of multiple usability evaluation approaches, not all the usability evaluation method types are suited for individuals with impairments. For example, Kashimoto et al. [106] apply the Wizard of Oz method with unstructured interviews during the iterative usability test for the development of a smart glass prototype. During the evaluation, Kashimoto et al. [106] find it was difficult for older adults with dementia to stay in a stationary position for a long time, focus, and accomplish the navigational task, which resulted in the researchers gathering unsatisfactory data for further analysis. However, the qualitative data were gathered through interviews. All the methods are also not suitable for participants with a minimal education level. For example, Sin et al. [107] apply four usability constructs that is, ease of use, efficiency, effectiveness, and user satisfaction, to evaluate their prototype. While collecting the data to measure the system's usability through a system usability scale (SUS), the researchers find the participant's had a lack of formal education, and as a result, all the questions had to be read by "the researcher, and they only were asked to point their answer on a graphical Likert scale" (p.124).

5 Findings and Recommendations

In addition to usability issues, the current study revealed additional findings, especially regarding the (i) social-technical aspects, (ii) effect of usability, and (iii) individual preferences of wearables. For example, the study conducted by Bower and Sturman [108] finds that participants were concerned about the privacy of people taking their photos and recording videos of them. Although, an interaction technique facilitates the user experience, similarly, Ye et al. [74] show that the interaction technique also brings social-technical challenges. For example, in their study, participants had difficulties using speech as the interaction techniques in a public environment because of privacy concerns. Similarly, in the study of Wulf et al. [81], some of the older adults were concerned about using speech interaction in public and felt uncomfortable doing so. We also found that a barrier, such as a design flaw on the physical design of the device or implementation bug on the user interface, causes frustration, fatigue among individuals with impairments at faster rate than with individuals without impairments [62]. Similarly, in another study conducted by Wulf et al. [81], the emotions of the users' reactions changed when the speech interaction did not function properly. Angelini et al. [109] also find that older adults have different preferences of the devices, stating, "The medical feature should not emerge in the product and should be presented as an accessory feature, which, however, will be appreciated by their relatives" (p.431). Similarly, Ye et al. [74] show that the aesthetics aspect of the device is relatively important for visually impaired participants.

Although we provided a comprehensive overview of the usability issues through a categorization framework and category summary (see Table 4), the literature review shows that these identified issues are still unsolved and need immediate attention from technology designers, researchers, and application developers. In this section, we discuss some of the obstacles, including design; individual preferences; device usage; and data, that are causing the identified usability issues and have been discussed and need further investigation. Furthermore, these challenges still persist because some aspects of the characteristics of wearable devices, such as the wearing position, do not satisfy an individual's daily hedonic or utilitarian (practical) needs, affecting their emotions, personal taste, self-expressive dimension (social and altruistic value), aesthetic, and functionally related values [110–112].

The SLR shows that background of wearable devices is becoming increasingly heterogeneous because of the rapid rise of (i) several categories of wearable devices, such as smartwatches, pedometers, implants, and HMDs and (ii) an embracing of the culture for monitoring, tracking, delivering, augmenting, and assisting purposes in both one's personal and work environment [1]. In addition, one of the insights gained from the SLR is that the usability issues surrounding the user interface, product aesthetics which can degrade user performance and user dissatisfaction [113], are related to the attributes of the user's characteristics, such as age and the user's background [74, 114]. Designing the user interface (i.e., visual interface and non-visual wearable interface) and the product aesthetics for wearability, accessibility, and readability which fulfills attributes of user's characteristics poses trade-off challenges to both device manufacturer and the application developers. Although many researchers try to offset these issues through with user-centric design approaches [67, 80] or by applying universal design principles and guidelines [15, 23, 115, 116], challenges still persist. Gandy et al. [15] state, "making devices that all individual can access at all times isn't always possible" (p.19). For example, we believe that these challenges persist because the design guidelines are usually created based on the individual's characteristics, such as age and disability, without looking at the individual's daily different use contexts and the device form factors. Kim et al. [117] state that, "Different usability problems are experienced more often according to different use contexts" (p.9). Because wearable devices can be utilized in various use contexts within the work environment and home and because they have versatile input systems in various form factors, including smart clothing, ring, necklace, wristband, and on the body [118], it is critical for research community to find ways to overcome future challenges.

Although one of the major goals of wearables that depend on sensors is to mediate the experience of reality between the individual and data and develop an intimate relationship between them [119], when either one or multiple wearables are connected to a single hub, that is, an external device, and the information is delivered using a single user interface on the external devices, a challenge arises regarding the data quality standard (i.e., availability, usability, reliability, relevance, and presentation quality) [120]. For example, the SLR reveals that the data quality standard is obstructed by suboptimal app crashes; poor synchronization of data because of a lack of connectivity; and motion artifacts. Looking closer at this challenge, the impact of data quality will have a significant implication for user experiences [121], more negative and arousing subjective feelings of excessive self-monitoring, a false sense of security, or may fuel a self-driven misdiagnosis [10] that can demoralize users' emotions, ability, and motivation, impacting the success of the applications. For example, Fogg [122] asserts that for behavior to happen, a person must have sufficient motivation, sufficient ability, and an effective trigger. Similarly, Zadra and Clore [123] point out that emotions can routinely alter the perceptions of individuals, here stating, "positive moods encourage one to maintain one's current way of looking at things, and that negative moods encourage a change" (p.10). To improve the quality of the data from the app, utilizing the early prediction approach for application crashes presented by Xia et al. [121], which is based on a naive Bayes model, before releasing the apps to the individuals should be considered. Similarly, the research community and technology designers should consider looking for techniques to improve the usability challenges regarding the poor synchronization of the data. One way is to use future complementary wireless networking techniques, such as "light fidelity (Li-Fi)" or "data through illumination (D-light)," which both provide additional free and vast wireless capacity, along with the ability to enhance the spectrum efficiency of existing radio frequency (RF) networks [124]. Although Li-Fi requires light to pass through the device, this technique could be implemented in wearable devices such as smartwatches and pedometers, which have a user interface and could easily interact with the light for data transmission, hence improving communication, speed, flexibility, and usability [125]. Similarly, another way to improve the quality of the quantified data is by reducing the impact of the time discrepancies in the data itself, which usually occur during the data fusion from multiple wearable devices converging with the external devices. Here, either model presented by Xu et al. [126] can be applied: a single-modal normal distribution (SMND) model for devices in which the data are generated with static frequency, for example, heart rate data, or a multi-modal normal distribution (MMND) model for devices in which data are generated with a dynamic frequency, for example, the step data collected by a smartwatch. Although we discussed improving the data by reducing the time discrepancies and using complementary wireless technology, looking closer, the main issue still remains when it comes to motion artifacts because of predetermined and orientation positioning relative to the individual's body [100], sources of nuisance, such as measurement noise [127], and failure to recognize activity by the sensors. We suggest that technology designers and researchers should consider utilizing the K-nearest neighbor (KNN) and its ensemble classification method with a proper choice of key parameters. This will have significant impact on the recognition accuracy when it comes to designing a robust and responsive machine learning in the wearables, as described by [40].

Beyond the user interface and data challenges, the additional major challenges that the SLR shows with regards to current device manufacturers face in understanding the individual preferences and device usage such as which sort of device shape and size, material do individual prefer; preferred position to wear the device, are those devices utilized for tracking physical activities, adopted as fashion accessories or used for educational or entertainment purposes. Abbate et al. [96] assert, "Ergonomic and aesthetic modifications are necessary to improve the level of usability and acceptability" (p.231). One way to move forward is to improve the design of the product's aesthetics, making them unique in their design, keeping them lightweight, choosing materials that depend less on the internal components and instead using a modular-based approach where individuals can change the device's external look and feel based on their daily needs. The question is then how to design a device that is more lightweight than what is available now. One possibility could be reducing the battery size and increasing battery life by (i) utilizing the work presented by Shen et al. [128], which is based on graphene-based supercapacitor fabrics with a high energy density and load-bearing capability or by (ii) harvesting energy from alternative sources, such as heat and motion from the body in the form of kinetic or thermal energy [129]. In addition, implementing these techniques would not only improve the weight, but also reduce the charging inconvenience [130], improving wearing behavior and ultimately leading to long-term use.

In the future, when more wearable devices such as HMDs are utilized for daily use purposes, more challenges with the visual interface design will appear, especially with the "output," that is, how much information will be delivered to the user and for which type of devices. For example, delivering information on the screen display of a HMD may not be the same as delivering information on smartwatches or external devices because information on a HMD will create a visual distraction and further complexity. In addition to delivering information, other challenges are the usage modes with wearable devices, the user interface, and the wearable's associated applications on external devices. For example, when multiple wearable devices become part of the user's daily life, the user will have different usage modes, such as sequential usage (i.e., moving from one device to another at different times to accomplish a task) or simultaneous usage (i.e., using more than one device at the same time for either a related or an unrelated activity) [131]. This could cause challenges related to usability, learnability, effectiveness, efficiency, memorability, errors, user satisfaction, task-technology fit, accessibility, orientation clues, conciseness, and cognitive load [132–134].

To conclude, the categorization framework and category summary (see Table 4) show that the usability challenges related to wearable devices are well known in the human–computer interaction (HCI) field; however, one could argue that these challenges still persist because the research community within the HCI field is focused on identifying and solving problems by conducting usability evaluations, using less targeted participants who are within a specific geographic location, rather than understanding the emotions and perceptions of larger groups of individuals, applications from a demographic context (e.g., age, gender, impairments, education level, employment status, and culture) [135]. Moreover, device manufacturers (i) are more focused on horizontal innovation, which solely implies changes in the current product characteristics; and vertical innovation, where new additional features are implemented, or the technical characteristics are improved to compete with other device manufacturers [136]; (ii) sponsored validation tests which may not be independently verified and may be difficult to understand and replicate individually [70]. Additionally, application developers are more focused on developing applications without understanding the users' needs. For example, commercial off-the-shelf (COTS) wearable devices are mostly designed by the designer, who is located in a specific location, who does not have knowledge of other culture, and who simply localized user interface based on the translation. Similarly, application developers create an application in one geographic location and target it to different locations through the application store. Although this allows device manufacturers and application developers to release their products to be used everywhere, sometimes with support and special features for use in a specific locale [137], issues arise from the cultural context, including the "anthropological culture," "symbolic culture," and "culture as community" [138]. Smith and Yetim [139] state, "Effective strategies that address cultural issues in both the product and the process of information systems development" now often are critical to systems success" (p.2). In considering how to increase the adoption of wearables by individuals, future research should incorporate greater variations of larger groups of individuals to analyze their emotions, and perceptions toward existing wearable devices within a certain demographic context (e.g., age, gender, impairments, education level, employment status, and culture) [135].

On the other hand, most of the reviewed papers perform usability evaluations with devices that are in the prototype phase or with devices that are already available on the market. Usability evaluations throughout the design cycle of product development are critical to ensure that the products are usable. There is currently no usability evaluation method for detecting and mapping usability issues from the initial stage of the development process of wearable devices to their release. Because wearable devices need to be reliable and wearable, the traceability of the usability evaluation together with users from different demographic contexts is crucial during each further stage of the development process to identify user needs and eliminate usability issues. Further research should be oriented toward identifying possible usability evaluation methods and integrating effective usability evaluations into the wearable development process. The results obtained from each usability evaluation can thus be effectively evaluated to ensure the reliability of the wearable devices to create commercially viable devices. There are different categories of wearable devices, as well as usability evaluation methods.

The limitations of the current study relate to its reliance on the articles from previous research and the inclusion and exclusion criteria. In addition, the search keywords were limited to "usability issues" when "wearable devices" was the targeted keyword. However, access to relevant papers depends on the precision of the search strings. Because the current study only concentrated on the above-listed search keywords, it is possible that other relevant articles could have been retrieved with different sets of keywords. The search also produced non-relevant articles and low-quality publications that were ignored. In addition, the main search was conducted only in English, which limits the results and maximizes bias.

6 Conclusion

In different product forms, wearable devices are appearing rapidly in the market, but the usability of them is challenging. Although many studies conduct a usability evaluation, a comprehensive overview on which types of usability issues currently exist for which types of devices is lacking. The uniqueness of the current work is in it filling this gap by identifying, analyzing, and providing a comprehensive overview of current trends of usability issues found in relevant studies. The review has revealed that in current research, usability challenges related to wearable devices can be categorized into device characteristics, deployment on the body, and the external devices used to synch with the wearables. In many cases, the usability issues are caused by a halo effect within device characteristics or device characteristics and wearing position. For example, data inaccuracy is caused by motion artifacts or by device connectivity.

Overall, the proposed categorization framework and category summary (see Table 4), of the usability issues generated from prior studies shows researchers, practitioners, and application developers in the wearable domain what challenges they have to consider to improve the design of the various types of wearable devices. Although the presented categorization framework, and category summary (see Table 4) provides an overview, there are still challenges that must be overcome in terms of design; individual preferences [140]; device usage; and data, all of which are causing the identified usability issues. Improving these open challenges will likely improve the adoption of wearable devices, however requiring to strengthen coordination between researchers, practitioners, and application developers. Additionally, the current study identified the most frequently used evaluation methods (i.e., types and classes, automation, and simulations) utilized for measuring the usability of wearable devices. It was found that experiments, surveys and questionnaires, and interviews were the most employed UEMs type and that inquiry was the most common UEM class. Moreover, the summary (see Table 5) provides an overview that can be used by practitioners and application developers to understand and make decisions while selecting the UEM for a particular type of device evaluation.

The present study can, however, be used as a basis for further studies to (i) extend new usability issues for upcoming wearable devices; (ii) discover how a categorization framework of usability issues varies across different demographics (i.e., age, culture, and gender); (iii) quantitatively identify the predominant usability issues from the proposed categorization framework; and (iv) extend usability evaluation method for new type of wearable devices.

References

1. Khakurel, J., Melkas, H., & Porras, J. (2018). Tapping into the wearable device revolution in the work environment: A systematic review. *Information Technology and People, 31*, 791–818. https://doi.org/10.1108/ITP-03-2017-0076.
2. Motti, V. G., & Caine, K. (2014). Human factors considerations in the design of wearable devices. *Proceedings of the Human Factors and Ergonomics Society Annual Meeting, 58*, 1820–1824. https://doi.org/10.1177/1541931214581381.
3. Rapp, A., & Cena, F. (2016). Personal informatics for everyday life: How users without prior self-tracking experience engage with personal data. *International Journal of Human Computer Studies, 94*, 1–17. https://doi.org/10.1016/j.ijhcs.2016.05.006.
4. Lee, J., Kim, D., Ryoo, H.-Y., & Shin, B.-S. (2016). Sustainable wearables: Wearable technology for enhancing the quality of human life. *Sustainability, 8*, 466. https://doi.org/10.3390/su8050466.
5. Clawson, J., Pater, J. A., Miller, A. D., et al. (2015). No longer wearing: Investigating the abandonment of personal health-tracking technologies on craigslist. In *Proceedings of the 2015 ACM International Joint Conference on Pervasive and Ubiquitous Computing – UbiComp '15* (pp. 647–658). New York, NY: ACM Press.
6. Lazar, A., Koehler, C., Tanenbaum, J., & Nguyen, D. H. (2015). Why we use and abandon smart devices. In *Proc eedings of the 2015 ACM International Joint Conference on Pervasive and Ubiquitous Computing – UbiComp '15* (pp. 635–646). https://doi.org/10.1145/2750858.2804288.
7. Endeavor Partners. (2014). *Inside wearables – Part 2*.
8. Blevis, E. (2007). Sustainable interaction design: Invention & disposal, renewal & reuse. In *Conference on Human Factors in Computing Systems – CHI '07* (p. 503). https://doi.org/10.1145/1240624.1240705.
9. Lin, Y., Breugelmans, J., Iversen, M., & Schmidt, D. (2017). An Adaptive Interface Design (AID) for enhanced computer accessibility and rehabilitation. *International Journal of Human Computer Studies, 98*, 14–23. https://doi.org/10.1016/j.ijhcs.2016.09.012.
10. Piwek, L., Ellis, D. A., Andrews, S., & Joinson, A. (2016). The rise of consumer health wearables: Promises and barriers. *PLoS Medicine, 13*. https://doi.org/10.1371/journal.pmed.1001953.
11. Abbas, S. Q. (2010). Development of a quality design framework for usable User Interfaces. *International Journal of Computational Science and Engineering, 02*, 1763–1767.
12. Trivedi, M. C. (2012). Role of context in usability evaluations: A review. *Advanced Computing: An International Journal, 3*, 69–78. https://doi.org/10.5121/acij.2012.3208.
13. ISO. (2009). ISO 9241-210:2009. *Ergonomics of human system interaction – Part 210: Human-centred design for interactive systems*.
14. Gafni, R. (2009). Usability issues in mobile-wireless information systems. *Issues in Informing Science and Information Technology, 6*, 754–769.
15. Gandy, M., Ross, D., & Starner, T. E. (2003). Universal design: Lessons for wearable computing. *IEEE Pervasive Computing, 2*, 19–23. https://doi.org/10.1109/MPRV.2003.1228523.
16. Liu, X., Vega, K., Maes, P., & Paradiso, J. A. (2016). Wearability factors for skin interfaces. In *Proceedings of the 7th Augmented Human International Conference 2016 on – AH '16* (pp. 1–8). New York, NY: ACM Press.
17. Stone D, Jarrett C, Woodroffe M, Minocha S (2005) User interface design and evaluation, 1st. Interactive technologies. Morgan Kaufmann. 704 pages. ISBN-10: 0120884364, ISBN-13:978-0120884360.
18. Petersen, K., Feldt, R., Mujtaba, S., & Mattsson, M. (2008). Systematic mapping studies in software engineering. In *EASE'08: Proceedings of the 12th International Conference on Evaluation and Assessment in Software Engineering* (pp. 68–77). https://doi.org/10.1142/S0218194007003112.

19. Steiger, E., de Albuquerque, J. P., & Zipf, A. (2015). An advanced systematic literature review on spatiotemporal analyses of twitter data. *Transactions in GIS, 19*, 809–834.
20. Alves, V., Niu, N., Alves, C., & Valença, G. (2010). Requirements engineering for software product lines: A systematic literature review. *Information and Software Technology, 52*, 806–820.
21. Akhavian, R., & Behzadan, A. (2016). Wearable sensor-based activity recognition for data-driven simulation of construction workers' activities. In *Proceedings - Winter Simulation Conference* (pp. 3333–3344).
22. Rapp, A., & Cena, F. (2015). *Affordances for self-tracking wearable devices* (pp. 141–142). ISWC. https://doi.org/10.1145/2802083.2802090.
23. Tomberg, V., Scholz, T., & Kelle, S. (2015). *Universal access in human-computer interaction. Access to interaction.* Cham: Springer.
24. Dhawale, P. P., & Wellington, R. J. (2015). Identifying the characteristics of usability that encourage prolonged use of an Activity Monitor. In *New Zealand Conference on Human-Computer Interaction* (pp. 39–42). https://doi.org/10.1145/2808047.2808056.
25. Jiang, H., Chen, X., Zhang, S., et al. (2015). Software for wearable devices: Challenges and opportunities. In *Proceedings – International Computer Software and Applications Conference* (pp. 592–597).
26. Kitchenham, B., & Charters, S. (2007). *Guidelines for performing systematic literature reviews in software engineering.* Technical report, Keele University and University of Durham.
27. Engström, E., & Runeson, P. (2011). Software product line testing – A systematic mapping study. *Information and Software Technology, 53*, 2–13. https://doi.org/10.1016/j.infsof.2010.05.011.
28. Budgen, D., & Brereton, P. (2006). Performing systematic literature reviews in software engineering. In *Proceeding of the 28th International Conference on Software Engineering – ICSE '06* (p. 1051). New York, NY: ACM Press.
29. Tosi, D., & Morasca, S. (2015). Supporting the semi-automatic semantic annotation of web services: A systematic literature review. *Information and Software Technology, 61*, 16–32.
30. Petticrew, M., & Roberts, H. (2006). *Systematic reviews in the social sciences: A practical guide.* Wiley-Blackwell. ISBN: 978-1-405-12110-1
31. Kitchenham, B., Pretorius, R., Budgen, D., et al. (2010). Systematic literature reviews in software engineering – A tertiary study. *Information and Software Technology, 52*, 792–805. https://doi.org/10.1016/j.infsof.2010.03.006.
32. Welsh, E. (2002). Dealing with data: Using NVivo in the qualitative data analysis process. *Forum Qualitative Social Research, 3*, Art 26. https://doi.org/10.17169/fqs-3.2.865.
33. NVIVO. (2018). *NVIVO.* http://www.qsrinternational.com/nvivo/what-is-nvivo. Accessed 1 Feb 2018.
34. Elsevier. (2008). *Mendeley.* https://www.mendeley.com/
35. Ozkan, B. C. (2004). Using NVivo to analyze qualitative classroom data on constructivist learning. *Environments, 9*, 589–603.
36. Ivory, M. Y., & Hearst, M. A. (2001). The state of the art in automating usability evaluation of user interfaces. *ACM Computing Surveys, 33*, 470–516. https://doi.org/10.1145/503112.503114.
37. Kaewkannate, K., & Kim, S. (2016). A comparison of wearable fitness devices. *BMC Public Health.* https://doi.org/10.1186/s12889-016-3059-0.
38. Impellizzeri, F. M., & Bizzini, M. (2012). Systematic review and meta-analysis: A primer. *International Journal of Sports Physical Therapy, 7*, 493–503.
39. Nowell, L. S., Norris, J. M., White, D. E., & Moules, N. J. (2017). Thematic analysis: Striving to meet the trustworthiness criteria. *International Journal of Qualitative Methods, 16.* https://doi.org/10.1177/1609406917733847.
40. Janidarmian, M., Fekr, A. R., Radecka, K., & Zilic, Z. (2017). A comprehensive analysis on wearable acceleration sensors in human activity recognition. *Sensors (Switzerland), 17.* https://doi.org/10.3390/s17030529.

41. McHugh, M. L. (2012). Interrater reliability: The kappa statistic. *Biochemia Medica*, 276–282. https://doi.org/10.11613/BM.2012.031.
42. Hosseini, M., Shahri, A., Phalp, K., et al. (2015). Crowdsourcing: A taxonomy and systematic mapping study. *Computer Science Review, 17*, 43–69. https://doi.org/10.1016/j.cosrev.2015.05.001.
43. Ally, M., & Gardiner, M. (2012). Application and device characteristics as drivers for smart mobile device adoption and productivity. *International Journal of Organizational Behavior, 17*, 35–47.
44. Liu, C.-C., Wu, D.-W., Jou, M., & Tsai, S.-J. (2010). Development of a sensor network system for industrial technology education. In *Communications in Computer and Information Science* (pp. 369–374).
45. Dennis A, Wixom BH, Roth RM (2012) Systems analysis and design, 6th ed. Wiley. (October 29, 2014). November 3 2014, ASIN: B00P6SS8OG https://www.amazon.com/Systems-Analysis-Design-Alan-Dennis-ebook/dp/B00P6SS8OG.
46. Leinonen, T., Purrna, J., Ngua, K., & Hayes, A. (2013). Scenarios for peer-to-peer learning in construction with emerging forms of collaborative computing. In *International Symposium on Technology and Society, Proceedings* (pp. 59–71).
47. Rauschnabel, P. A., Hein, D. W. E., He, J., et al. (2016). Fashion or technology? A fashnology perspective on the perception and adoption of augmented reality smart glasses. *i-com, 15*. https://doi.org/10.1515/icom-2016-0021.
48. Oh, S., So, H., & Gaydos, M. (2017). Hybrid augmented reality for participatory learning: The hidden efficacy of the multi-user game-based simulation. *IEEE Transactions on Learning Technologies, 11*, 1. https://doi.org/10.1109/TLT.2017.2750673.
49. Wichrowski, M., Koržinek, D., & Szklanny, K. (2015). Google glass development in practice. In *Proceedings of the Mulitimedia, Interaction, Design and Innnovation on ZZZ – MIDI '15* (pp. 1–12). New York, NY: ACM Press.
50. Kim, K. J. (2017). Shape and size matter for smartwatches: Effects of screen shape, screen size, and presentation mode in wearable communication. *Journal of Computer Communication, 22*, 124–140. https://doi.org/10.1111/jcc4.12186.
51. Pulli, P., Hyry, J., Pouke, M., & Yamamoto, G. (2012). User interaction in smart ambient environment targeted for senior citizen. *Medical & Biological Engineering & Computing, 50*, 1119–1126. https://doi.org/10.1007/s11517-012-0906-8.
52. Harrington, C., Wood, R., Breuer, J., et al. (2011). Using a unified usability framework to dramatically improve the usability of an EMR module. *American Medical Informatics Association Annual Symposium Proceedings, 2011*, 549–558.
53. Laarni, J., Heinilä, J., Häkkinen, J., et al. (2009). Supporting situation awareness in demanding operating environments through wearable user interfaces. In *Lect Notes Comput Sci (including Subser Lect Notes Artif Intell Lect Notes Bioinformatics) 5639 LNAI* (pp. 13–21). https://doi.org/10.1007/978-3-642-02728-4_2.
54. Delabrida, S. E., DAngelo, T., Oliveira, R. A. R., & Loureiro, A. A. F. (2015). Building wearables for geology. In *2015 Brazilian Symposium on Computing Systems Engineering (SBESC)* (pp. 148–153). IEEE.
55. Laramee, B., Laramee, R. S., & Ware, C. (2002). Rivalry and interference with a head mounted display. *ACM Transactions on Computer-Human Interaction, 9*, 238–251. https://doi.org/10.1145/568513.568516.
56. Laramee, R. S., & Ware, C. (2001). Visual interference with a transparent head mounted display. In *CHI '01 extended abstracts on Human factors in computer systems – CHI '01* (p. 323). New York, NY: ACM Press.
57. McGill, M., Murray-Smith, R., Boland, D., & Brewster, S. A. (2015). A dose of reality. In *Proceedings of the 33rd Annual ACM Conference Extended Abstracts on Human Factors in Computing Systems – CHI EA '15* (pp. 177–177). New York, NY: ACM Press.
58. Shih, P. C., Han, K., Poole, E. S., et al. (2015). *Use and adoption challenges of wearable activity trackers*. iConference Proc 1–12.

59. Holzinger, A., Searle, G., Pruckner, S., et al. (2010). Perceived usefulness among elderly people: Experiences and lessons learned during the evaluation of a wrist device. In *Procceedings of the 4th International ICST Conference on Pervasive Computer Technology and Healthcare* (pp. 1–5). https://doi.org/10.4108/ICST.PERVASIVEHEALTH2010.8912.
60. Costanza, E., Inverso, S., & Pavlov, E. (2006). eye-q: Eyeglass peripheral display for subtle intimate notifications. In *Proc 8th* ... (pp. 211–218). https://doi.org/10.1145/1152215.1152261.
61. Jacob, R. J. K. (2000). New human-computer interaction techniques. In J. Hyona, & R. Radach (Eds.), *The mind's eye: Cognitive and applied aspects of eye movement research.* Elsevier Science BV. ISBN: 0–444–51020–6.
62. Neto, L. d. S. B., Maike, L. V. R. M., et al. (2015). A wearable face recognition system built into a smartwatch and the blind and low vision users. *Lecture Notes in Business Information Processing, 241,* 515–528. https://doi.org/10.1007/978-3-319-29133-8.
63. Lawo, M., Herzog, O., & Witt, H. (2007). An industrial case study on wearable computing applications. In *Proceedings of the 9th International Conference on Human Computer Interaction with Mobile Devices and Services – MobileHCI '07* (pp. 448–451). https://doi.org/10.1145/1377909.1378052.
64. Han, T., Han, Q., Annett, M., et al. Frictio: Passive kinesthetic force feedback for smart ring output. In *Proceedings of the UIST 2017* (Vol. 2017).
65. Brun, D., Ferreira, S. M., Gouin-Vallerand, C., & George, S. (2016). CARTON project: Do-it-yourself approach to turn a smartphone into a smart eyewear. In *Proceedings of the 14th International Conference on Advances in Mobile Computing & Multimedia* (pp. 128–136). https://doi.org/10.1145/3007120.3007134.
66. "Claire", L. S., & Starner, T. (2010). BuzzWear: Alert perception in wearable tactile displays on the wrist. In *Proceedings of the 28th International Conference on Human Factors in Computing Systems – CHI '10* (pp. 433–442). https://doi.org/10.1145/1753326.1753392.
67. Levin-Sagi, M., Pasher, E., Carlsson, V., et al. (2007). A comprehensive human factors analysis of wearable computers supporting a hospital ward round. *IEEE Xplore,* 1–12.
68. Manabe, H., & Fukumoto, M. (2011). Tap control for headphones without sensors. In *Proceedings of the 24th Annual ACM Symposium on User Interface Software and Technology – UIST '11* (p. 309). https://doi.org/10.1145/2047196.2047236.
69. Zhang, Y., & Rau, P. P. (2015). Playing with multiple wearable devices: Exploring the influence of display, motion and gender. *Computers in Human Behavior, 50,* 148–158. https://doi.org/10.1016/j.chb.2015.04.004.
70. Grant, A. (2017). Is my data valid? In: *QS labs.* http://quantifiedself.com/2017/08/validating-self-tracking-devices/. Accessed 30 Mar 2018.
71. Rodríguez, I., Cajamarca, G., Herskovic, V., et al. (2017). Helping elderly users report pain levels: A study of user experience with mobile and wearable interfaces. *Mobile Information Systems, 2017,* 1–12. https://doi.org/10.1155/2017/9302328.
72. Ross, D. A., & Blasch, B. B. (2002). Development of a wearable computer orientation system. *Personal and Ubiquitous Computing, 6,* 49–63. https://doi.org/10.1007/s007790200005.
73. Rasche, P., Wille, M., Theis, S., et al. (2015). Activity tracker and elderly. In *Computing Information Technology Ubiquitous Computing Communications Dependable, Auton Secur Comput Pervasive Intell Comput (CIT/IUCC/DASC/PICOM), 2015 IEEE International Conference on IS – SN – VO – VL* (pp. 1411–1416). https://doi.org/10.1109/CIT/IUCC/DASC/PICOM.2015.211.
74. Ye, H., Malu, M., Oh, U., & Findlater, L. (2014). Current and future mobile and wearable device use by people with visual impairments. *Chi, 2014,* 3123–3132. https://doi.org/10.1145/2556288.2557085.
75. Altenhoff, B., Vaigneur, H., & Caine, K. (2015). One step forward, two steps back: The key to wearables in the field is the app. In *Proceedings of the 9th International Conference on Pervasive Computing Technologies for Healthcare* (pp. 241–244). https://doi.org/10.4108/icst.pervasivehealth.2015.259049.

76. Ananthanarayan, S., Sheh, M., Chien, A., et al. (2014). Designing wearable interfaces for knee rehabilitation. In *Proceedings of the 8th International Conference on Pervasive Computing Technologies for Healthcare*. ICST.

77. Ferri, J., Lidón-Roger, J. V., Moreno, J., et al. (2017). A wearable textile 2D touchpad sensor based on screen-printing technology. *Materials (Basel), 10.* https://doi.org/10.3390/ma10121450.

78. Savindu, H. P., Iroshan, K. A., Panangala, C. D., et al. (2017). BrailleBand: Blind support haptic wearable band for communication using braille language. In *Proceedings of the IEEE International Conference on Systems, Man and Cybernetics* (pp. 1381–1386). Canada: Cybern Banff Center, Banff. https://doi.org/10.1109/SMC.2017.8122806.

79. Chen, S., Lan, Y.-C., Zheng, Y.-R., et al. (2015). Usability of a low-cost wearable health device for physical activity and sleep duration in healthy adults. In *Proceedings of the 2015 Work Pervasive Wireless Healthcare – MobileHealth '15* (pp. 35–38). https://doi.org/10.1145/2757290.2757298.

80. Thorpe, J. R., Rønn-Andersen, K. V. H., Bień, P., et al. (2016). Pervasive assistive technology for people with dementia: A UCD case. *Healthcare Technology Letters, 3,* 297–302. https://doi.org/10.1049/htl.2016.0057.

81. Wulf, L., Garschall, M., Himmelsbach, J., & Tscheligi, M. (2014). Hands free – Care free. In *Proceedings of the 8th Nordic Conference on Human-Computer Interaction Fun, Fast, Foundational – NordiCHI '14* (pp. 203–206). New York, NY: ACM Press.

82. Carter, S., Marlow, J., Komori, A., & Mäkelä, V. (2016). Bringing mobile into meetings: Enhancing distributed meeting participation on smartwatches and mobile phones. In *Proceedings of the 8th International Conference on Human-Computer Interaction with Mobile Devices and Services* (pp. 407–417). https://doi.org/10.1145/2935334.2935355.

83. Ahanathapillai, V., Amor, J. D., & James, C. J. (2015). Assistive technology to monitor activity, health and wellbeing in old age: The wrist wearable unit in the USEFIL project. *Technology and Disability, 27,* 17–29. https://doi.org/10.3233/TAD-150425.

84. Sultan, N. (2015). Reflective thoughts on the potential and challenges of wearable technology for healthcare provision and medical education. *International Journal of Information Management, 35,* 521–526. https://doi.org/10.1016/j.ijinfomgt.2015.04.010.

85. Yang, P., Hanneghan, M., Qi, J., et al. (2015). Improving the validity of lifelogging physical activity measures in an internet of things environment. In *2015 IEEE International Conference on Computer and Information Technology; Ubiquitous Computing and Communications; Dependable, Autonomic and Secure Computing; Pervasive Intelligence and Computing* (pp. 2309–2314). IEEE.

86. Koskimäki, H., Mönttinen, H., Siirtola, P., et al. (2017). Early detection of migraine attacks based on wearable sensors. In *Proceedings of the 2017 ACM International Joint Conference on Pervasive and Ubiquitous Computing and Proceedings of the 2017 ACM International Symposium on Wearable Computer – UbiComp '17* (pp. 506–511). https://doi.org/10.1145/3123024.3124434.

87. Albrecht, U. V., Von Jan, U., Kuebler, J., et al. (2014). Google glass for documentation of medical findings: Evaluation in forensic medicine. *Journal of Medical Internet Research, 16.* https://doi.org/10.2196/jmir.3225.

88. Rahmati, A., Qian, A., & Zhong, L. (2007). Understanding human-battery interaction on mobile phones. In *Proceedings of the 9th International Conference on Human Computer Interaction with Mobile Devices and Services – MobileHCI '07* (pp. 265–272). New York, NY: ACM Press.

89. Brewster, S., Lumsden, J., Bell, M., et al. (2003). Multimodal "eyes-free" interaction techniques for wearable devices. In *Proceedings of the Conference on Human Factors in Computing Systems – CHI '03* (p. 473).

90. Spagnolli, A., Guardigli, E., Orso, V., et al. (2014). Measuring user acceptance of wearable symbiotic devices: Validation study across application scenarios. In *Lecture Notes in Computer Science (LNCS), including its subseries Lecture Notes in Artificial Intelligence (LNAI) and Lecture Notes in Bioinformatics (LNBI)* (Vol. 8820, pp. 87–98). https://doi.org/10.1007/978-3-319-13500-7_7.

91. Oakley, I., Sunwoo, J., & Cho, I.-Y. (2008). Pointing with fingers, hands and arms for wearable computing. In *Proc CHI Ext Abstr* (pp. 3255–3260). https://doi.org/10.1145/1358628.1358840.
92. Ju, A. L., & Spasojevic, M. (2015). Smart jewelry. In *Proceedings of the 2015 Workshop on Future Mobile User Interfaces – FutureMobileUI '15* (pp. 13–15). New York, NY: ACM Press.
93. Goto, M., Kimata, H., Toyoshi, M., et al. (2015). A wearable action support system for business use by context-aware computing based on web schedule. In *Proceedings of the 2017 ACM International Joint Conference on Pervasive and Ubiquitous Computing and Proceedings of the 2017 ACM International Symposium on Wearable Computer – UbiComp '15* (pp. 53–56). https://doi.org/10.1145/2800835.2800863.
94. Kondo, Y., Takahashi, S., & Tanaka, J. (2015). Information select and transfer between touch panel and wearable devices using human body communication. In M. Kurosu (Ed.), *Lecture Notes in Computer Science (including subseries Lecture Notes in Artificial Intelligence and Lecture Notes in Bioinformatics)* (pp. 208–216). Cham: Springer.
95. Nirjon, S., Gummeson, J., Gelb, D., & Kim, K.-H. (2015). TypingRing: A wearable ring platform for text input. In *Proceedings of the 13th Annual International Conference on Mobile Systems, Applications, and Services – MobiSys '15* (pp. 227–239). https://doi.org/10.1145/2742647.2742665.
96. Abbate, S., Avvenuti, M., & Light, J. (2014). Usability Study of a wireless monitoring system among Alzheimer's disease elderly population. *International Journal of Telemedicine and Applications, 2014*. https://doi.org/10.1155/2014/617495.
97. Fang, Y.-M., & Chang, C.-C. (2016). Users' psychological perception and perceived readability of wearable devices for elderly people. *Behaviour and Information Technology, 35*, 225–232. https://doi.org/10.1080/0144929X.2015.1114145.
98. Mizuno, T., & Kume, Y. (2014). Development of a glasses-like wearable device to measure nasal skin temperature. *Communications in Computer and Information Science, 435*, 338–342. https://doi.org/10.1007/978-3-319-07854-0.
99. Yoo, J., Nockhwan, K., Jeongho, K., & Hwan, R. J. (2014). Preliminary guidelines to build a wearable health monitoring system for patients: Focusing on a wearable device with a wig. *Communications in Computer and Information Science, 435*, 338–342. https://doi.org/10.1007/978-3-319-07854-0.
100. Yurtman, A., & Barshan, B. (2017). Activity recognition invariant to sensor orientation with wearable motion sensors. *Sensors (Switzerland), 17*. https://doi.org/10.3390/s17081838.
101. Klingeberg, T., & Schilling, M. (2012). Mobile wearable device for long term monitoring of vital signs. *Computer Methods and Programs in Biomedicine, 106*, 89–96. https://doi.org/10.1016/j.cmpb.2011.12.009.
102. Liang, Z., Nagata, Y., Martell, M. A. C., & Nishimura, T. (2016). Nurturing wearable and mHealth technologies for self-care: Mindset, tool set and skill set. In *2016 IEEE 18th International Conference on e-Health Networking, Applications and Services (Healthcom) 2016* (pp. 6–10). https://doi.org/10.1109/HealthCom.2016.7749432.
103. Masai, K., Kunze, K., Sugiura, Y., et al. (2017). Evaluation of facial expression recognition by a smart eyewear for facial direction changes, repeatability, and positional drift. *ACM Transactions on Interactive Intelligent Systems, 7*, 1–23. https://doi.org/10.1145/3012941.
104. Young, K. A. (2005). Direct from the source: The value of "think-aloud" data in understanding learning. *Journal of Educational Enquiry, 6*, 19–33.
105. Hafiz, P., Miskowiak, K. W., Kessing, L. V., et al. (2019). The internet-based cognitive assessment tool: System design and feasibility study. *JMIR Formative Research, 3*, e13898. https://doi.org/10.2196/13898.
106. Kashimoto, Y., Firouzian, A., Asghar, Z., et al. (2016). *Twinkle Megane: Near-eye LED indicators on glasses in tele-guidance for elderly.* 2016 IEEE International Conference on Pervasive Computing and Communication Workshops (PerCom Workshops).

107. Sin, A. K., Zaman, H. B., Ahmad, A., & Sulaiman, R. (2015). Evaluation of wearable device for the elderly (W-Emas). In H. Badioze Zaman, P. Robinson, A. F. Smeaton, et al. (Eds.), *Lecture Notes in Computer Science (including subseries Lecture Notes in Artificial Intelligence and Lecture Notes in Bioinformatics)* (pp. 119–131). Cham: Springer. https://link.springer.com/10.1007/978-3-319-25939-0_11.

108. Bower, M., & Sturman, D. (2015). What are the educational affordances of wearable technologies? *Computers in Education, 88*, 343–353. https://doi.org/10.1016/j.compedu.2015.07.013.

109. Angelini, L., Caon, M., Carrino, S., et al. (2013). Designing a desirable smart bracelet for older adults. In *Proceedings of the 2013 ACM Conference on Pervasive and Ubiquitous Computing Adjunct Publication – UbiComp '13 Adjun* (pp. 425–434). https://doi.org/10.1145/2494091.2495974.

110. Kumar, M., & Noble, C. H. (2016). Beyond form and function: Why do consumers value product design? *Journal of Business Research, 69*, 613–620. https://doi.org/10.1016/j.jbusres.2015.05.017.

111. Mugge, R., & Schoormans, J. P. L. (2012). Product design and apparent usability. The influence of novelty in product appearance. *Applied Ergonomics, 43*, 1081–1088. https://doi.org/10.1016/j.apergo.2012.03.009.

112. Reinecke, K., & Bernstein, A. (2011). Improving performance, perceived usability, and aesthetics with culturally adaptive user interfaces. *ACM Transactions on Computer-Human Interaction, 18*, 1–29. https://doi.org/10.1145/1970378.1970382.

113. Hinckley, K., Pausch, R., Goble, J. C., & Kassell, N. F. (1994). A survey of design issues in spatial input. In *Proceedings of the 7th Annual ACM Symposium on User Interface Software and Technology – UIST '94* (pp. 213–222). New York, NY: ACM Press.

114. Hwang, S., Song, J., & Gim, J. (2015). Harmonious haptics: Enhanced tactile feedback using a mobile and a wearable device. In *Ext Abstr ACM CHI'15 Conference on Human Factors in Computing Systems* (Vol. 2, pp. 295–298). https://doi.org/10.1145/2702613.2725428.

115. Tomberg, V., & Kelle, S. (2018). *Universal design based evaluation framework for design of wearables* (pp. 105–116).

116. Wentzel, D. J., Velleman, E. M., & van der, G. D. T. (2016). Wearables for all: development of guidelines to stimulate accessible wearable technology design. In *13th Web for All Conference. Part 34*.

117. Kim, H., Kim, J., Lee, Y., et al. (2002). An empirical study of the use contexts and usability problems in mobile Internet. In *Proceedings of the Annual Hawaii International Conference on System Sciences* (pp. 1767–1776).

118. Yoon, S. H., Huo, K., & Ramani, K. (2016). Wearable textile input device with multimodal sensing for eyes-free mobile interaction during daily activities. *Pervasive and Mobile Computing, 33*, 17–31. https://doi.org/10.1016/j.pmcj.2016.04.008.

119. Swan, M. (2015). Connected car: Quantified self becomes quantified car. *Journal of Sensor and Actuator Networks, 4*, 2–29. https://doi.org/10.3390/jsan4010002.

120. Cai, L., & Zhu, Y. (2015). The challenges of data quality and data quality assessment in the big data era. *Data Science Journal, 14*, 2. https://doi.org/10.5334/dsj-2015-002.

121. Xia, X., Shihab, E., Kamei, Y., et al. (2016). Predicting crashing releases of mobile applications. In *Proceedings of the 10th ACM/IEEE International Symposium on Empirical Software Engineering and Measurement – ESEM '16* (pp. 1–10).

122. Fogg, B. J. (2003). Persuasive technology: Using computers to change what we think and do. *Persuas Technol Using Comput to Chang What We Think Do, 5*, 283. https://doi.org/10.4017/gt.2006.05.01.009.00.

123. Zadra, J. R., & Clore, G. L. (2011). Emotion and perception: The role of affective information. *Wiley Interdisciplinary Reviews: Cognitive Science, 2*, 676–685. https://doi.org/10.1002/wcs.147.

124. Haas, H., Yin, L., Wang, Y., & Chen, C. (2016). What is LiFi? *Journal of Lightwave Technology, 34*, 1533–1544. https://doi.org/10.1109/JLT.2015.2510021.

125. Sharma, V., Rajput, S., & Sharma, P. K. (2016). *Light fidelity (Li-Fi): An effective solution for data transmission* (p. 020061).

126. Xu, T., Guo, A., Ma, J., & Wang, K. I.-K. (2017). Feature-based temporal statistical Modeling of data streams from multiple wearable devices. In *2017 IEEE 15th International Conference on Dependable, Autonomic and Secure Computing, 15th International Conference on Pervasive Intelligence and Computing, 3rd International Conference on Big Data Intelligence and Computing and Cyber Science and Technology Congress(DASC/PiCom/DataCom/CyberSciTech)* (pp. 119–126). IEEE.
127. Cenedese, A., Susto, G. A., & Terzi, M. (2017). A parsimonious approach for activity recognition with wearable devices: An application to cross-country skiing. In *2016 European Control Conference, ECC 2016* (pp. 2541–2546).
128. Shen, C., Xie, Y., Zhu, B., et al. (2017). Wearable woven supercapacitor fabrics with high energy density and load-bearing capability. *Scientific Reports, 7*. https://doi.org/10.1038/s41598-017-14854-3.
129. Hannan, M. A., Mutashar, S., Samad, S. A., & Hussain, A. (2014). Energy harvesting for the implantable biomedical devices: Issues and challenges. *Biomedical Engineering Online, 13*.
130. Jeong, H., Kim, H., Kim, R., et al. (2017). Smartwatch wearing behavior analysis. *Proceedings of the ACM Interactive, Mobile, Wearable Ubiquitous Technology, 1*, 1–31. https://doi.org/10.1145/3131892.
131. Google Inc. (2012). *The new multi-screen world study*. Google Think Insights.
132. Ismail, N. A., Ahmad, F., Kamaruddin, N. A., & Ibrahim, R. (2016). A review on usability issues in mobile applications. *IOSR Journal of Mobile Computing & Application, 3*, 47–52. https://doi.org/10.9790/0050-03034752.
133. Ito, K., Sugano, S., Takeuchi, R., et al. (2013). Usability and performance of a wearable tele-echography robot for focused assessment of trauma using sonography. *Medical Engineering & Physics, 35*, 165–171. https://doi.org/10.1016/j.medengphy.2012.04.011.
134. Nielsen, J. (2012). *Usability 101: Introduction to usability*. https://www.nngroup.com/articles/usability-101-introduction-to-usability/
135. Papadopoulos, K. (2014). The impact of individual characteristics in self-esteem and locus of control of young adults with visual impairments. *Research in Developmental Disabilities, 35*, 671–675. https://doi.org/10.1016/j.ridd.2013.12.009.
136. Cecere, G., Corrocher, N., & Battaglia, R. D. (2015). Innovation and competition in the smartphone industry: Is there a dominant design? *Telecommunications Policy, 39*, 162–175. https://doi.org/10.1016/j.telpol.2014.07.002.
137. Ayyal Awwad, A. (2017). Localization to bidirectional languages for a visual programming environment on smartphones. *International Journal of Computer Science Issues, 14*, 1–13. https://doi.org/10.20943/01201703.113.
138. Collins, R. (1990). *Culture, communication and national identity: The case of Canadian television*. Toronto: University of Toronto Press.
139. Smith, A., & Yetim, F. (2004). Global human-computer systems: Cultural determinants of usability. *Interacting with Computers, 16*, 1–5. https://doi.org/10.1016/j.intcom.2003.11.001.
140. Khakurel, J., Porras, J., & Melkas, H. (2019). Human-centered design components in spiral model to improve mobility of older adults. In S. Paiva (Ed.), *Mobile solutions and their usefulness in everyday life* (pp. 83–104). Cham: Springer.

Mobile-Based Indian Currency Detection Model for the Visually Impaired

Abhishek Pathak and Sagaya Aurelia

1 Introduction

In India, the crime rate is increasing at an alarming rate. It is common to hear about scams all the time; every day you'll find some innocent getting scammed for money. Money is an important aspect of everyday life as nothing is free, but the limits that people go to for attaining money is getting worse day by day. It is hard to read and not think about what would be going on with the people suffering specially if they are not physically or mentally fit. Like people try to scam people with visual impairment by fooling them in way that utilizes the persons weakness, in this instance visual impairment. The victim doesn't even realize it until it's too late, and now there is nothing he could possibly do. If we consider Indian currency, it's poorly designed with respect to a visually impaired person as there is no provision provided for judging the value of the currency, specially after the demonetization of old notes of 500 and 1000 in India. The new currency notes were so similar in size with all the other notes that the visually impaired people started it find it a lot challenging as they were easily fooled by scammers who used to give them smaller notes promising them to be of a higher value. For example, if the visually impaired user took a cab and wants to pay the cab driver for his service and realizes that he doesn't have any change – in this case the visually impaired will give him a note of higher value and would expect some cash in return; although, if the cab driver was a scammer, the cab driver could give even smaller note than he is expected to give or would deny that the previous note was the same amount that was expected to be paid originally [1–2].

A. Pathak (✉) · S. Aurelia
Department of Computer Science, CHRIST (Deemed to be University), Bengaluru, India
e-mail: abhishek.pathak@mca.christuniversity.in; sagaya.aurelia@christuniversity.in

© Springer Nature Switzerland AG 2020
S. Paiva, S. Paul (eds.), *Convergence of ICT and Smart Devices for Emerging Applications*, EAI/Springer Innovations in Communication and Computing,
https://doi.org/10.1007/978-3-030-41368-2_3

Table 1 Input to output
comparison of currency
detection model

Input	Output
2000 rupees	4 vibrations and audio
500 rupees	3 vibrations and audio
200 rupees	2 vibrations and audio
100 rupees	1 vibration and audio
Any other valid currency note	Only audio
Others	Only audio

So, we are introducing the mobile-based Indian currency detection model which would enable a visually impaired to check the value of a given currency. The mobile-based Indian currency detection model is a mobile-based application where the user clicks a picture of the currency note he is holding, then the data is sent to the online server for preprocessing [3]. After receiving the data, which is apparently an image, the server would process the data and give a suitable output to the mobile device of the user. The mobile device would then receive the output, and depending upon what output was received from the server, the device would manipulate it to the user. The output would be in the form of audio and vibrations so the user gets to know the output even if the mobile device is on silent mode, as shown in Table 1 [4].

The model is mobile based because we wanted it to be easy to carry and user-friendly as possible; to make this possible, various facts and things were kept into consideration while designing the model. The main factors taken into consideration while designing this model were

(a) As visually impaired-friendly as possible
(b) Should be supported for all mobile devices
(c) Should be fast and accurate
(d) Both the faces of the notes are same
(e) Provision of reporting a wrong result

(a) As visually-impaired friendly as possible:
    The model is being designed for a visually impaired person so it is kept in mind that it supports audio output. This would allow the user to understand what is going on in his mobile device. The application would be kept short and to the point with just the provision of clicking and picture of a currency and getting the value of the currency [5].
(b) Should be supported for all mobile devices:
    Mobile devices come with various configurations to specifically design the model to support all the different devices where specifically low-end devices would be hassle, as we won't be knowing how much memory they support and if it would be enough for running our model. So, we are implementing an online server to run our model, while the mobile device would be used to click an image for the required resolution and send it to the online server. This would not be possible without an Internet connection that's why its compulsory for the user to have an Internet connection enabled in his mobile device.

Fig. 1 Showing two faces of the same note

(c) Should be fast and accurate:

The other reason the model is being hosted online and not kept inside the user's mobile device is because we want faster processing speed and we wouldn't want users with low-end configuration devices to be stuck and keep on waiting for a possible result. Hosting the server would also allow the convolutional neural network to get a bigger dataset and keep on improving. Also, the user can report any wrong outputs, and we would have the original image and the output, and we can analyze what went wrong with it.

(d) Both the faces of the notes are same:

As you can see in Fig. 1, a visually impaired person doesn't have any idea whether they are holding the note front up or front down, so it's a bit of challenged, and the design keeps in mind that both of the given conditions should be treated as a valid currency note and even specify the user about the not being in face-down position so that the user can turn over the note and reverify the previous result and get a much better surety of the value of the currency.

(e) Provision of reporting a wrong result:

The model would also have a provision to allow the visually impaired person to report a result if in case the output was wrong and he got to know about it from some other sources. This would allow the convolutional neural network to adapt to any and would improve its chances of getting wrong answers in future.

2 Proposed Model

The mobile-based Indian currency detection model is designed to be used in a mobile device. In this model the mobile device would be used for taking a picture of a person's hand holding a currency note and then the image would be sent on the server where there will be basic steps taken for processing of the image and for image segmentation. For this to happen, an Internet connection is needed; otherwise the image won't be processed. After receiving the image on the web server, the server would start the preprocessing of the image so as to identify the currency and to make the image ready to be sent to the neural network. While processing the image, we would also make sure to resize the image so that it fits inside our neural network perfectly. After doing all the preprocessing, we would like to send the data to the convolutional neural network which would then check the input image and tell the outcome by passing it over all its layers on neurons. The neural network we are using is of a classifier type, which means that it would take an input, and after processing the input, it would classify the input into one of the predefined categories. For example, we are going to work on an image of a currency so the network will take the image and classify it into what the value of the specific note is that the user is holding in his hand.

The basic structure of the model can be seen in Fig. 2 where we can see that the image would be sent for preprocessing and after the image is processed, then we would try to segment the currency out of the user's hand. After we have a separated the image of the note, we could send it to the convolutional neural network for classification. This would enable the neural network to only learn the features that are actually related to features in an Indian currency and not something from the background details. This could also possibly help us later on in determining the difference between a counterfeit currency and an original currency note as the neural network is trained on useful stuff and not on any extra features like hand color or features related to a human palm.

2.1 Acquiring Image

The first step in mobile-based Indian currency detection model is the acquiring of the image, and this is done by the mobile device's camera, as depicted in Fig. 3,

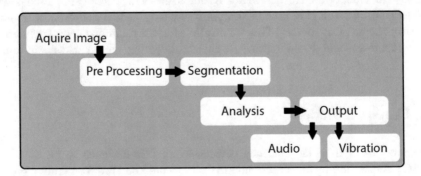

Fig. 2 Flow of currency detection model

Fig. 3 A sample input image for the model

where the user would be allowed to click a picture of the note he holding. This would be done by the user opening the application and pointing the camera toward his other hand with the money and clicking an image. The whole application would be voice-based navigation so as to make it friendly to users who are suffering from visual impairment. The image would then be sent to the online server for further preprocessing and checking, and the application would wait for a response from the server; in case no response is received, the application would notify the user that something went wrong or maybe the server is down. This would allow the user to always be informed of any scenario that happens [6, 7].

2.2 Preprocessing

This step helps the model to ready the image for segmentation. Basically, it's performing basic steps to change the image into something from which we can easily identify what is what. As the input image would contain only three major features, a hand, a note, and a background, to further classify the value of the currency, the image of the note should be classified first. So, to recognize the value of the currency, we need the image of the note extracted out of user's hand and then only we can perform any other classification techniques on the image. Extracting the note out of the hand could become a hectic task, as if you go for features like trying to extract the biggest rectangle out of the original image. It would fail as in real-life scenarios, the corners could have been folded or just hidden under a figure, hence generalizing the scenario and proceeding didn't help. So, proceeding with taking the color as the major differentiating point between the three was much more feasible as all of the three are usually distinct and it provided much better results. For the classification a K means segmentation technique is used where we take the average of the major colors present in the image and all the colors that are similar to our different averages; we replace the close-by color to the average color [8].

Fig. 4 Extraction of a note
using k mean algorithm

The number of averages to be taken is taken as an input in K means algorithm; for the required model, we took it as 3 as it depicts the number of unique items of the image, and we have to just find the average color of the note inside the image, and we could then easily classify our image from that. This was easily done by taking a small circle at the center of the image and calculating which of the major colors had the greatest number of pixels in the center of the image. As noticed that the note is at the focus of the image, it usually occupies the maximum space or is at the center of the image; after realizing the correct channel of the image, we can easily find the biggest closed figure made by that color (which obviously is the note) and pass it for further segmentation [9]. There are more steps being performed while performing this preprocessing in a step-by-step way. Like, for example, we would first resize the image so that all the images we get are of the same size each and use a Gaussian blur filter which would allow us to blur out sharp edges and details. Normally, high-resolution cameras tend to take a sharper image where a lot of minute details would be mistaken for edges; blurring the image would basically smudge out all the details, and only a kind of average color of the whole thing would be visible which would then help us separate objects based on their color. Then we would apply the K means segmentation algorithm, where we would specify the color's channel count as 3. Basically, this would allow to break the picture down in three different parts; then we would extract the biggest and the centermost object in the whole of image as shown in Fig. 4. This is then passed on to the next step of the model where the real segmentation of the image would take place [10, 11].

2.3 Segmentation

This step deals with the real separation of the note out of the original image, to extract the identified image of the currency out from the rest of the background by using the K mean sample sent by after being preprocessed. This is easily performed

Fig. 5 Note correctly identified

by comparing the original image after resizing and finding the coordinates of the K mean segmented image; that is why the image was resized and was kept common at the start only. As now both the K mean segmented image and the original image are of same size, a pixel of the same value in both of the images represents the same thing. Now simply find all the pixels that enclose the image provided by the K mean segmented image, and use them to enclose the note in the original image [12]. Once this step is done, this means that we have successfully identified the note inside the image. This step could be easily seen in Fig. 5 where the user is holding the same note, but now we know where the note is inside this image; after knowing that much we can easily extract out the image of the note by either copying all the values inside the enclosed box into a new empty image or either emptying out all the pixels that are not enclosed by the box. The extracted-out image would be then again resized as per the requirement for our convolutional neural network and then converted into black and white as we do not require colors for checking the value of the currency. This would allow us to save space and computational power as a colored image is treated as three images in computer with red, green, and blue images stacked upon each other, while gray-scale images are simply shades of gray. This helps us in sending neural network inputs as it would take only integer inputs, and if we used a color image, we would have to create three times the input layer than we are going to require in just a black and white image meaning it's working like a row of numbers. So, when we are trying to pass a two-dimensional sequence of number, we flatten them into one single long row and send them to the neural network for analysis, unless we are dealing with a colored image [13]. A colored image is not just a sequence of two-dimensional number, but instead it could be treated as three images of the same thing with different colors, one with shades of red, the other two with blue and green, respectively. Hence, when attempting to send a colored image to a neural network, we would have to perform the same process of flattering a two-dimensional image into a single row of numbers, but this time we have to continue this same process thrice to convert all the three color channel images into single rows and then send a single row that is three times the size of the row when taken in black and white case. This means that taking the same thing in as a colored image would simply use up more resources and would even require more computational

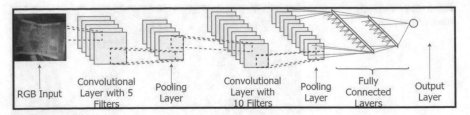

RGB Input | Convolutional Layer with 5 Filters | Pooling Layer | Convolutional Layer with 10 Filters | Pooling Layer | Fully Connected Layers | Output Layer

Fig. 6 Basic representation of a neural network

power. While dealing with a server-based model, we not only consider the accuracy of the model, but also the speed and passing of black and white image would allow us much faster speed of producing a result with the same accuracy [14, 15].

2.4 Analysis

The segmented image is then extracted and converted into black and white then sent to the convolutional neural network for analyzing the value of the note extracted. This simply works on the basis of imaging the whole image in a two-dimensional matrix, and each container contains a pixel value which determines the color of the image in shades of gray where 1 is black and 0 is white. So each of the pixels in the image is an input for the neural network as depicted in Fig. 6 [16].

The whole input layer would consist of the values of the pixels flatted into a single row of number where each and every pixel value represents the color of that pixel in numbers, like, for example, if we have a pixel value of 1, it's black; 0 is white, but a pixel value of 0.5 would be considered to be gray; likewise, a value of 0.25 is light-colored gray, and 0.75 is dark-colored gray. The other different layers inside a neural network beside the input and output layer are known as hidden layers. A hidden layer is a layer whose exact working we don't know as it changes every time the model predicts wrong output and it readjusts itself for the correct input [17, 18].

Two major things to consider while working with neural networks are

- Bias
    A bias represents what is the neural network more inclined to do when stuck in a tuff spot, meaning if the neural network is 50% sure it's a 2000-rupee note and 50% sure that the object is not a valid note, what will the network choose as both the results are equally valid? Thus, for cases like this, a bias value is passed to neural network that helps it decide what option to make when both the options are equally likely. For our mobile-based Indian currency detection model, we are considering not a valid currency as a bias. Meaning if the neural network is stuck in deciding a single option at any time, it would automatically consider that the currency is not valid.

- Selection function

    This is another major concept to take care as selection function represents at what condition should it signal the next layer to proceed. Meaning it is the condition that if satisfied, the neural network would say yes to the next layer. As neural network works with number and there is always some integer value in each of the node of the neural network for like is the network is 70% sure that the given currency is a 2000-rupee note, that means it's 30% sure that it's some other note than a 2000-rupee note. Hence, a selection fitness determines when the node is fit enough to pass a confirmation output to the next node which would then check its fitness and trigger its next note. This keeps on happening in the whole network with the input of each and every input. For this current model, we are using a sigmoid function which basically deals with values between 1 and 0, with 1 being yes and other being no [19].

Thus, after all the nodes are checked, the final probability is passed to the last layer which is the output layer. This layer basically determines what exactly is the output. Thus, it contains nodes with each and every possible outcome including a node for the note being not valid. At last whichever node has the highest value is the output of the neural network, and its value represents the accuracy of the result. Like if the node of 500 rupee has a value of 0.6 at the last node, that means the model is 60% sure the input image had a note of 500 rupee in it. The maximum valued node after calculation is the answer, which is then sent back to the device from where the input came.

2.5   Output

Once we receive back a reply signal from the online server, we would then proceed with convening the results to the user. This would be done in processing of the received output by the mobile device and itself knowing about the result. The received signal would be checked, and depending upon the value of the reply, we will proceed on telling the user the output in the required format. The output is produced in audio, vibration, or in both formats, which depends upon what was the result of the analysis. Table 1 shows us a better view of whatever the possible output could be for a given picture. The vibration feature was added to the system for helping the user even if his mobile's audio is turned off; hence we use it for big amounts of money so that it doesn't get scammed.

3   Result and Discussion

The discussion was done as a proposal for mobile-based Indian currency detection model. There are multiple things that could be improved like while working with low-quality cameras, this happens due to the usage of a Gaussian filter over an already blurred image as it would blur the details even more and make the edges even harder to detect. Although if we didn't use Gaussian filter at all, it would just make it much harder as the main motive of using a Gaussian filter is used to remove out the tiny details that come in because of using a highly pixelated camera. The better the camera quality, the better the edges inside an image, and it would create chaos during image segmentation. In this model the image segmentation works on basically splitting the given image into three basic parts an outer background, a hand, and a note, so it's not necessary for us to check for all those minute details on the currency. The other factor that contributes to the degradation of result is a 10-rupee note which has almost the same color as skin colors. This was the major challenge we faced during the implementation of this model as some people have almost the same skin color as of the 10-rupee note; otherwise, an easier option could have been to black out all the skin-colored pixels and the we would have been left with a note and a side background of the image. Normal images, where we just have a single object over a background, could be easily detected with using contours and its properties. But in implementing the model for real-life usage and scenarios, we are supposed to provide more accurate results. Hence, we are using an approach of extracting the image out of the hand separately and then trying to learn its features.

4   Limitation

The mobile-based Indian currency detection model works on the simple basis of extracting the centermost contour, meaning it would find the biggest shape in the center of the image. We are doing this with the help of K means clustering. K means clustering allows us to extra three major components of the image using their color, the note, hand, and background. After the K means clustering algorithm, we simply extract the centermost distinct object based on the color. This helps us in segmenting the note in a better way as the note isn't always in rectangular shape and could possibly have folded edges or some portion hidden by the hand. Hence, it's no use to look for a big rectangle in the image. So, for the model to work, it is necessary for the note to be at the center of the image. If not, the model will not work and provide an unexpected output which is not good. In real-life scenarios, it is not certain for the image to be perfect as the user is someone with visual impairment, and it could happen that the user didn't realize that the image was imperfect and would blame the other person for scamming. There have been special measures taken to help the mobile-based Indian currency detection model detect the correct currency part of the note.

Another major limitation for this model to work is that it needs Internet connectivity to the online server to run the model which has a limitation of its own, like the server needs to be fast in processing and responding. In case of the server being too busy or the server being down, a visually impaired user won't be able to verify his currency. So, the runtime of the server and the performance of the server also play a keen factor in determining the outcome and should be of good quality, although the user needs to have a proper Internet connection which is not possible at every possible location. Thus, a module for offline classification should be considered, but then the mobile device should be able to process all the data at a very fast rate, which would not be possible in allowing configuration of device. Counterfeit currency is also a major problem as the current model would not be able to recognize counterfeit currency, which very closely resembles that of an original note but with some very minute distinct features.

5 Future Enhancements

The mobile-based Indian currency detection model only works in recognizing whether there is a currency note inside an image and recognizing the exactly the value for it, but it could be easily fooled with a replica of the original currency. Visually impaired people are smart enough to recognize whether the paper that is handed to them is in fact from an Indian currency note or not, but a close replica can fool them. A fully abled person with full eye vision could get fooled if he's not paying attention, so it's fairly obvious a close replica would be hard to detect; even banks use UV rays and other technologies that could not be embedded inside a mobile device that's why it's hard to detect a counterfeit currency from an original currency accurately. Although we shall try to increase the accuracy as much as possible by implementing different image processing algorithms once we are sure that the given image is a valid currency note of a specific value, which would be the output of our current model. Future improvements will also include a neural network just for checking if what is found is an original currency or a counterfeit currency by comparing the various symbols and security markings provided inside the currency by the Indian government.

Currently the model is designed to handle only a single currency at a time. The model works properly, but it takes too much time in the case of need, as a person could be holding more than one note at a particular instance of time. Thus, it would take too long for a person to check each and every currency note one by one, which not only is a waste of time for the user but also for the person who is offering the currency in exchange. As someone who is trying to attempt a scam won't simply stand there and wait until for his deeds to get caught, the scammer would possibly try to hurry things up and wouldn't allow the visually impaired user to check each and every currency note in the mobile-based Indian currency detection model. Thus, this would make the model ineffective in real-life scenarios as it's too slow for usage at an instance like this, but in future implementation, we shall try to enable it to work for multiple currency notes at a single instance of time. This would allow a visually impaired person to not have to wait for checking and wait for a response from the server one by one, but instead multiple notes can be clicked at once, and the system will automatically detect the different notes checked for the original values and notify the user separately about each one of them. We know that Internet connectivity is not possible at every possible location. Hence an offline implementation of the model would also be provided. Although the offline variant would only be supported on high-end mobile devices only, it would be better not to implement it until the neural network has an accuracy of 95% at least. This would allow a better and faster alternative system in the future which would not require connectivity to the Internet.

6   Conclusion

The main motive for the creation of the mobile-based Indian currency detection model was for helping the visually impaired people and giving them a chance of defending themselves from scammers who try to take profit out of their disability. As scamming and cheating are a punishable offense and people should be discouraged not to do it, trying to scam someone who is specially abled is just inhuman and makes us realize to what extent humanity has fallen. Thus, for the proposed model, we are using OpenCV and TensorFlow library in Python; OpenCV is an image processing library which allows us to do image processing. We used OpenCV to ready the image before sending it to the neural network, while TensorFlow is an Google-provided deep learning library which revolutionized the deep learning industry. TensorFlow with the help of Keras is used to host deep learning models where TensorFlow is the back end used for all the processing and running of the model, while Keras is the front end of the model which is used to add layers and design the network. This model utilizes a simple convolutional neural network for image classification, while for hosting the model, a virtual hosting is required. There is no highly specific need for the hosting, but a fast hosting service that is able to run Python-based scripts and has a graphics processing unit would be great.

References

1. Guo, T., Dong, J., Li, H., & Gao, Y. (2017). Simple convolutional neural network on image classification. In *2017 IEEE 2nd international conference on Big Data Analysis (ICBDA)* (pp. 721–724). Beijing.
2. Rathee, N., Kadian, A., Sachdeva, R., Dalel, V., & Jaie, Y. (2016). Feature fusion for fake Indian currency detection. In *2016 3rd international conference on Computing for Sustainable Global Development* (pp. 1265–1270). New Delhi: INDIACom.
3. Darade, S. R., & Gidveer, G. R. (2016). Automatic recognition of fake Indian currency note. In *2016 international conference on Electrical Power and Energy Systems (ICEPES)* (pp. 290–294). Bhopal.
4. Pissaloux, E. E. (2002). A vision system design for blinds mobility assistance. In *Proceedings of the second joint 24th annual conference and the annual fall meeting of the Biomedical Engineering Society Engineering in Medicine and Biology* (Vol. 3, pp. 2349–2350). Houston, TX.

5. Takatori, N., Nojima, K., Matsumoto, M., Yanashima, K., & Magatani, K. (2006). Development of voice navigation system for the visually impaired by using IC tags. In *2006 international conference of the IEEE Engineering in Medicine and Biology Society* (pp. 5181–5184). New York, NY.
6. Hassanpour, H., Yaseri, A., & Ardeshiri, G. Feature extraction for paper currency recognition, 1-4244-0779-6/07/$20.00 ©2007 IEEE.
7. Takeda, F., & Omatu, S. (1995). High speed papercurrency recognition by neural networks. *IEEE Transaction on Neural Networks, 6*(1), 73–77.
8. Alfarras, M. (2012). Bahraini paper currency recognition. *Journal of Advanced Computer Science and Technology Research, 2*(2), 104–115.
9. Vishnu, R., & Omman, B. (2014). Principal features for Indian currency recognition. In *Annual IEEE India conference*.
10. Bhavani, R., Karthikeyan, A., & Novel, A. (2014, April). Method for banknote recognition system. *IJCSE, 2*(4), 165–167.
11. Rajan, G. V., Panicker, D. M., Chacko, N. E., Mohan, J., & Kavitha, V. K. (2018). An extensive study on currency recognition system using image processing. In *2018 conference on Emerging Devices and Smart Systems (ICEDSS)* (pp. 228–230). Tiruchengode.
12. Ballado, A. H., et al. (2015). Philippine currency paper bill counterfeit detection through image processing using Canny Edge Technology. In *2015 international conference on Humanoid, Nanotechnology, Information Technology, Communication and Control, Environment and Management (HNICEM)* (pp. 1–4). Cebu City.
13. Sarfraz, M. (2015). An intelligent paper currency recognition system. *Procedia Computer Science, 65*, 538–545.
14. Konecki, M., Ivković, N., & Kaniški, M. (2016). Making programming education more accessible for visually impaired. In *2016 39th international convention on Information and Communication Technology, Electronics and Microelectronics (MIPRO)* (pp. 887–890). Opatija.
15. Kulkarni, A., Wang, A., Urbina, L., Steinfeld, A., & Dias, B. (2016). Robotic assistance in indoor navigation for people who are blind. In *2016 11th ACM/IEEE international conference on Human-Robot Interaction (HRI)* (pp. 461–462). Christchurch.
16. Soysa, L., Lokuge, K., Wimalasundera, I., & De Silva, M. N. (2010). Enhancing learning for Visually Impaired with technology: MATHVIS. In *2010 international conference on Technology for Education* (pp. 228–229). Mumbai.
17. Rao, S. N., & Suraj, R. (2016). Smartphone-aided reconfigurable multi-device controller system for the visually challenged. In *2016 IEEE international conference on Computational Intelligence and Computing Research (ICCIC)* (pp. 1–4). Chennai.
18. Froneman, T., van den Heever, D., & Dellimore, K. (2017). Development of a wearable support system to aid the visually impaired in independent mobilization and navigation. In *2017 39th annual international conference of the IEEE Engineering in Medicine and Biology Society (EMBC)* (pp. 783–786). Seogwipo.
19. Alzubi, J., Nayyar, A., & Kumar, A. (2018, November). Machine learning from theory to algorithms: An overview. *Journal of Physics: Conference Series, 1142*(1), 012012. IOP Publishing.

Home Supporting Smart Systems for Elderly People

Eleni Boumpa and Athanasios Kakarountas

1 Introduction

The aging of the population is a problem that concerns the society worldwide as a global issue. The increase of the elderly population is a certainty that affects every society, market, and industry, pushing them to adapt to a new reality. On the other hand, the elderly themselves are seeking solutions for independent and supportive living in their preferred setting. Thus, the so-called Silver Economy was surfaced, which is related to supporting the well-being of older people, with the aim of health and social care monitoring, health services, and self-health management of older people [1]. In this chapter, the aim is to provide a survey for every researcher who wants to understand and research in the topic of support systems for the elderly in their home environment. Special effort was given to present in detail each work, in order to facilitate understanding for researchers the historical evolution of the research topic. Additionally, the associated technologies and the required technical background as well as the techniques for interacting and assisting the elderly are reported.

With the term *Smart Home*, we mean a home that has a communication network which interconnects its basic electrical devices for remote control and approach. For a home to become a smart one, it should have three basic features: (i) internal network (wired or wireless), (ii) gateway for system control and management, and (iii) the home's automation, that is, the interconnection of the home appliances/products with systems and services out of it [2]. Furthermore, it is required to integrate sensors in order to measure (sense) environmental parameters and others related

E. Boumpa (✉) · A. Kakarountas
Intelligent Systems Laboratory, University of Thessaly, Lamia, Greece
e-mail: eboumpa@uth.gr; kakarountas@uth.gr

© Springer Nature Switzerland AG 2020
S. Paiva, S. Paul (eds.), *Convergence of ICT and Smart Devices for Emerging Applications*, EAI/Springer Innovations in Communication and Computing, https://doi.org/10.1007/978-3-030-41368-2_4

to living, and a system to report on status (as defined by the system or the user), and actuators in order to react to events.

Apart from the basic electrical appliances, the evolution of technology has led to the introduction of several specially designed devices that have been developed for the further intelligence of the home. This kind of devices can be sensors, actuators, and embedded systems. These devices have some common attributes, such as:

- *Sense and Act*: Sense ambient parameters and act under certain conditions.
- *Memorize*: Store the collected data.
- *Process*: Process the collected data.
- *Communicate*: Transfer data to/from the network.

More specifically, when smart devices detect events or changes in the environment parameters, then data is stored, and the actuators are activated. Additionally, the gateway manages the communication between the smart devices over the network. The communication is achieved via different existing protocols, like Bluetooth, ZigBee, Wi-Fi, etc. The network can be (i) local (client-server model), (ii) on the cloud (Software as Service, SaaS; Software as Platform, SaaP; etc.), or (iii) combination of both (Edge Computing and Fog Computing). At last, a dedicated system platform delivers services, such as connectivity support, actuation services, data processing, device management application support, and solution provider services, about use cases and user benefits [3].

In addition, smart homes may be distinguished based on their characteristics, as:

1. The "Autonomous Home", containing smart objects and generally autonomous devices and objects that act in a smart way.
2. The "Network Home", containing interconnected smart objects via the network (wired or wireless) to exchange information between them.
3. The "Ubiquitous Home", which has internal and external networks for interaction and remote control of the systems, as well as access to services and information inside and outside the home.
4. The "Learning Home", which collects and records data from the residents' activity and uses them to predict their needs.
5. The "Attentive Home", which has the ability to continuously record the activity and location of people and objects for the residents' needs [4].

The desire of the elderly for Active and Assisted Living (AAL) has led to the introduction of *Smart Home for Elderly*. Smart Home for Elderly could be described as a subcategory of smart home that is mainly dedicated to meeting the needs of the elderly. Thus, this kind of smart home is equipped with technological tools that monitor, facilitate, and assist the independence of the elderly. It enhances independent living; provides the physical and mental health-care monitoring, safety, and security with non-intrusive way, and improves social contact of the elderly [5].

The purpose of such a smart home is providing to the elderly:

- Support and assistance in their daily life activities, as well as in emergency cases.
- Remote support and follow-up by their physicians.
- Remote monitoring by their relatives, as well as their relatives' notification in case of need.

The rest of the chapter is organized as follows: in Sect. 2 the overall progress of the research topic is presented in detail, and in Sect. 3 the challenges that need to be addressed by new researchers are reported. In Sect. 4 the state-of-the art works are discussed, as well as their contribution. In Sect. 5 the technical background that is required is offered in terms of sensors technology, communication protocols, and storage issues. In Sect. 6 the future trends are considered as they have been identified so far. Finally, the chapter concludes in Sect. 7 with an extended discussion on the presented material and specifically the knowledge on techniques and technologies that are used by the researchers, as well as the promising works to achieve inclusion of a support system in elders' home.

2 Previous Work

The overall progress, in the smart home commercial solutions and research works, has improved the everyday life of the smart home's residents. But more specifically, smart homes for elder people provide aid and care to them and extend their independent living. In this section, commercial solutions for smart homes, as well as various projects and applications that have been proposed throughout the world to meet the needs of the elderly, are presented.

2.1 Commercial Solutions for Smart Homes

Security services for a smart home are presented in [6]. These services include smart locks that provide alerts at the cell phone when a door opens, indoor cameras for two-way communication, doorbell cameras that permit the home's residents to check and respond to a person at the door either they are at home or not, and thermostats that adjust automatically home temperature to the preferences of the home's residents. Communication between all the smart devices of the home and its residents is achieved through an intuitive touchscreen dashboard and a smart home application and via Google's smart speakers. Thus, this solution can provide elderly people with safety and security, as well as personalized control of the conditions prevailing within their home.

In [7] the services of another commercial smart home are presented. In this smart home, the residents can control smart devices via voice control, and an application for smartphones and tablets is offered. Every connected device of the smart home can be controlled remotely. Thus, the control of home devices can provide monitoring, warnings, safety, and security for the elder residents of this smart home.

The smart home that is presented in [8] provides to its elder residents remote control on heating, remote monitoring of their home, measurement of the indoor and outdoor conditions of the home environment, and optimization of their comfort and well-being at home. All these are achieved with the use of smart thermostats, smart radiator valves, smart door and window sensors, smart indoor siren, smart video doorbell, smart smoke alarm, smart outdoor camera, smart indoor camera, smart home weather station, and smart indoor sensors for humidity, air quality, noise, and temperature measures.

The smart home's solution in [9] provides smart home services via different kinds of sensors and actuators that set up at home. These services are focused on safety and security, with the use of alarms and door entry systems, and improvement of the residents' well-being, via the monitoring of home environment conditions, and remotely lighting monitoring.

In [10], the presented smart home exploits smart devices, like actuators and embedded systems, via a single button and an application for smartphones and tablets, allowing the residents to get remote control through a wireless network. This smart home through the control of the lighting, blinds, switches, heating, and cameras provides to the older people well-being services, safety, and security.

We can notice that all the commercial solutions for smart homes cover partially and support some of the needs and demands of older people. However, these smart homes can provide comfort and autonomy to people of all ages. Therefore, it is an imperative need to create commercial solutions for smart homes specifically designed to meet the needs of the elderly. Smart homes will provide basic functions – services to support the previously mentioned, such as assisting people on their daily routine (i.e., reminders to run their daily program or reminders for their medication, etc.), assuring security and safety.

2.2   Research Projects

A smart home system for elderly that introduces the integration of sensor technologies and the development of a framework for acquisition, processing, and exchanging of extracted information from the residents' living habits is presented in [11]. The system's operation succeeds using various types of sensors. Furthermore, the smart home consists of an artificial intelligence (AI) expert system that correlates information from the sensors and makes decisions regarding the elderly status and their physiological condition.

Nourizadeh et al. [12] proposed a smart home for the patient-oriented distributed tele-home-care system. The system allows all the type of system's users to interact with it (e.g., health-care providers, doctors, caregivers, medical call centers, patients, and familiars). The proposed system offers to all abovementioned activity monitoring of the elderly (via the home automation and sensors networks) and tele-health-care services between the elderly, their doctors, and their family (via a user-friendly video-conference system) and additionally offers the ability to integrate with other home/health systems.

An experimental medical tele-surveillance system is proposed in [13]. The scope of this system is to extend the time of living of individuals at their home. The system consists of five levels, (i) the sensor level, (ii) the transmission level, (iii) the software level, (iv) the network level, and (v) the human level. In the sensor level, various measurements of parameters associated to the older person are collected. In the transmission level, there are two types of circuits; one for the transmission and one for receiving the data that have been collected by the sensors. In the software level, there is a computer application for the older people and their familiars for information acquisition, data recording and analysis. Additionally, at the same level, another application for the medical center is found, from which the medical staff has access to the display of the medication of the elderly. In the network level, the transmission of the information from the elderly's computer to the medical center and vice versa is implemented, and in the human level, the four user profiles are defined: the administrator, the physician, the familiar, and the elderly.

In [14] a U-Health smart home for the elderly is presented. The U-Health smart home focuses on the use of an autonomic computing system based on a knowledge-based (KB) architecture. Thus, the information that is gathered via the sensors (both biosensors and environmental) is send to the autonomic system and then transferred in the knowledge base, which finally makes decision, on its own, about the elderly's health (regardless it is needed from a caregiver or not).

Klack et al. [15] proposed a sensor-based floor for integration to home environments, to assist older people living independently at their home. A grid of piezoelectric sensors is embedded to the floor, allowing the detection of a user's position and some of the qualitative aspects of moving behaviors, like downfalls, etc. The structure can determine parameters like the position of the user within the room, the pose of the user (standing, sitting, laying), the weight of the user, the entrance/exit of the user to the room, and users' movement behavior like velocity of pace, the movement direction, and user identification.

AmIVital [16] is a platform for health and well-being for ambient intelligent (AmI) environment technological services for elderly, patients of chronic diseases, and dependent people. This platform is developing personal environment services and applications to control and improve health, life habits, and social state of the elderly, etc. The architecture of the platform consists of fixed and mobile gateways (functional, technological, and infrastructure services).

In [17] a wellness determination process of the elderly living independently in a smart monitoring home is proposed. The presented novel framework can verify the behavior of the elderly, such as the usage of appliances, activity recognition, and forecast levels. The prototype that has been developed is easy to install and maintain in the elderly's home, while it is stable to execute multiple tasks, like data collection and analysis in real time.

Vuegen et al. [18] presented the validation of a distributed acoustic sensor network that observes the daily living activities of the elderly at home. The sensors that were used for recording the day-to-day life activities were both audio and ultrasound receivers. The elderly's daily living activities that were successfully monitored were cooking and eating, reading, using of the laptop, vacuum cleaning, walking around, and watching TV.

The bioAssist [19] is a home care platform that provides monitoring and communication services between the patients and their health-care professionals, relatives, and friends. The platform's environment consists of (a) an application for smart devices that incorporates the functions for the communication with the cloud-based platform, with the sensors and the smartwatches, (b) a web-based application that interacts with all the platform users, (c) a help-desk application that is accessible by the health-care operators for receiving the emergency requests from the patient, and (d) a back-end platform that is a set of cloud-based services and components. Every patient uses a bio-signal recording smart device. His/her doctor can use the platform tools to get access on the patient's medical history, laboratory test results, medication, allergies, and health condition and also define a specific monitoring and treatment schedule.

UbiCare [20] is a low-cost Arduino-based AAL system. Its purpose is to support activity and fall detection and promote well-being for the elderly at home. The wireless sensor network that installed in the elderly's home is responsible for monitoring environmental parameters, like temperature, humidity, etc., and elderly's daily activities, like moving, sitting, sleeping, usage of electrical appliances, etc. Also, UbiCare system provides alarms, like SMS, emails, and voice calls, to elderly's caregivers (e.g., relatives and/or doctors).

An open, low-cost, IoT-enabled home monitoring solution is presented in [21]. This solution provides wireless collection and storage of bio-signals data from elderly; communication between elderly with relatives, carers, and doctors; and monitoring of the elderly health status with real-time communication via video conferencing and data sharing. The system consists of a Telecare Gateway, built on Raspberry Pi, that is, the communication gateway between the patient and the sensors, and a cloud-based communication platform that realizes the communication between the elderly and their individuals.

In [22], the architecture for a home monitoring system is proposed, which is based on sensors connected to the Internet through a Wi-Fi router. For this monitoring system, special sensor prototypes were developed. The sensor prototypes include passive infrared (PIR) sensors for monitoring the movements into the house, toilet sensor monitoring the toilet access, and magnetic contact detecting doors and windows closing/opening. The sensors' data are sent to an IoT platform, in the cloud. So, this work contributes to the elimination of the need for a dedicated home gateway to implement a home monitoring system.

Yang et al. [23] developed a localization method for indoor tracking humans' position based on PIR sensors, an accessibility map, and an A-star algorithm. This method can track the person in a house and provide a solution for a home-robot localization. The steps of the system's functionality are three. In the first step, the grid-based accessibility map reflects the preferences of human visiting and the physical layout of the home. In the second step, the PIR sensors provide an external rough position of the human, according to the grid-based accessibility map. And, in the third step, the A-star algorithm estimates the trajectory and reduces the errors, after the fusion of the PIR sensors data and grid-based accessible map.

A human physical activity recognition system based on data collected from smartphone sensors is proposed in [24]. Smartphone's sensors, such as accelerometer, gyroscope, and gravity sensor, via machine learning practices, are used to target elderly's tasks like walking, running, sitting, standing, ascending, and descending stairs.

3 Challenges

The most challenging issues in the implementation of IoT environments, such as the case of a smart home, are security and privacy. These two issues are a great concern to the scientific community, because of the ever increase of devices and objects that can be connected to the Internet (Internet of Things and the Internet of Everything). More generally, it is an issue that requires a great deal of sensitivity and attention from both the scientists and companies, as well as from the users themselves. In particular, smart homes have raised important concerns about the protection of people's privacy, as their services manage very sensitive and personal data.

Panwar et al. [25] recorded the basic specifications and requirements to turn a smart home as secure and private as possible. These specifications are presented and analyzed in Table 1.

Furthermore, several severe attacks to the networks threat the Wireless Sensor Networks (WSNs), which are used for implementing a smart home. In an IoT-based system using WSNs, the collected data must be securely and efficiently transmitted. So, a smart home requires explicitly defined security constraints to become secure and the destination nodes must be secured in data transmission.

Pirbhulal et al. [26] categorized all security constraints into two categories:

1. The network security that includes the secured localization, non-repudiation, availability, access control, trustworthiness, and authentication.

    Traditional network security principles also apply to the IoT domain. Authentication and access control mechanisms are required in order to prevent malicious access to installed devices. Additionally, other security properties such as non-repudiation or availability have to be served, in order to provide a more robust service. The main difference with respect to traditional computing is that resources of devices are significantly limited, while it is not always feasible to apply the same countermeasures in the constraint environment of an IoT system.

Table 1 Basic specifications and requirements for a smart home

Requirement	Description
Authentication	The confirmation of the identity of the participants in a communication *Means: Digital signatures, secure identity transmission, physically unclonable functions (PUF), and secure key*
Authorization	The Provision of rights to a certified user *Means: Fine-grained policy management and different access-control methods*
Confidentiality	The disclosure of the messages to an authorized user and hide them from an "opponent" *Means: Encryption, secret-sharing, traffic padding, zero-knowledge proofs, proof-of-knowledge, group signatures, pseudonym systems*
Integrity	The ensure of consistency and identity of messages *Means: Cryptographic hashing, watermarking, holographic proofs, multi-party computations, timestamping, nonce, and sequence numbers*
Availability	The guarantee of fair operation, resources, and services to the authorized user *Means: Anomaly detection, firewalls, and special communication hardware preventing external malicious traffic to reach the network*
Non-repudiation	The obstruction of the sender and/or the recipient to deny a message *Means: Digital signatures*
Accountability or Auditing	The process of keeping any sender's, recipient's, or network's action to verify the action in the future
Deniable communication	The authorization of a prover to plausibly deny that a protocol instance was ever executed, for which the same prover was an active participant
Unlinkability	The user has the ability to use multiple aliases, but it is not possible to connect two pseudonyms used by the same homeowner for two different services
Non-transferability	Only one user of the house is the administrator, who has unique privileges
Forward secrecy	The guarantee that a session key derived from a long-term public-private key pair is secure

2. The data security that includes confidentiality, privacy, integrity, and data freshness.

   IoT devices mainly process personal data of the subjects present at the installation location. Health measurements, location, or activity data are characteristic cases of personal data captured by IoT devices and processed by IoT systems. Handling such data requires confidentiality and integrity measures in order to not affect subjects' privacy and at the same time to provide a useful service. Cryptographic techniques shall be applied in order to ensure that no unauthorized party will access subjects' personal data and that data is immutable through the whole workflow of the IoT system.

4 State of the Art

In Sect. 2 several different home supporting projects for elderly people that have been proposed over the years were presented. The aim of all these projects is to support the elderly at home through different approaches and using different technologies and technological means. However, the state of the art of home supporting smart systems for elderly people seems to be moving toward ubiquitous and disappearing computing systems with physical interaction with the users.

The Sthenos project [27] develops a reliable system of environment awareness for human behavior recognition and performing to offer support to elderly and chronic patients at their home. This work aims to develop methodologies and tools to compose pervasive human-centered systems, to understand the human state in assistive environments, by using audiovisual and biological signals. This project proposed a successful way for human behavior recognition from audiovisual content, using the fusion of noninvasive sensors, tools for identification of affect and emotion from audiovisual data in a noninvasive way, tools to analyze neurophysiological data, and tools for dynamic multimodal human-machine interaction.

Maglogiannis ct al. [28] presented an electronic reminder system for smart devices, like Pebble smartwatch and Android devices. Its functionality is to notify the patients at the chosen time, with audio and visual alerts, to adhere to their treatment. The reminder notifications for the patients are pushed to their Pebble smartwatch. Also, the patients' relatives, physician, and/or pharmacist can use a web application to create or update reminders and monitoring the adherence of patients' treatment. The proposed system is a cloud-based service that consists of three software modules: (a) a web-based application for the creation of reminders and monitor patients' adherence; (b) an Android application for the patients, to manage the reminders; and (c) an application for the Pebble smartwatch, which handles the notifications that are received from the Android device.

In [29] a disappearing computing system for older adult-assisted living is presented. This system was designed based on commercial off-the-shelf (COTS) products. The innovation of this system is that modified home appliances with which the elderly are already familiar with, like TV and phone, are used as the human-computer interface (HCI). This is an open system that transforms typical home appliances and enables older people to communicate with their relatives, caregivers, and doctors through teleconference, to read the news, and to send and read their emails, just by selecting a dedicated TV channel. So, it succeeds in elderly's assisted living with low training and decreases the degree of their depression and loneliness.

A User Location Discovery (ULD) home system is proposed in [30]. The innovation of this system is that the user is not forced to carry a device. Also, the devices used for the system's implementation are limited to simple sensors and no devices like camera and microphone that affect directly the privacy of the user. When any of the sensors detects the user in the house, it sends the appropriate signal to the Home Gateway. Then, a User Location Discovery Context Broker in the Home Gateway processes all the information and stores the user's location information in the Context Information Repository of the Home Gateway. Finally, the Main

Context Broker decides about the service that will be offered to the user via the home communication network.

Bafhtiar et al. [31] examine the feasibility of a cloud-based voice recognition technology as the means of home care, with the use of clinical decision support. For the support of home care, a custom application, with the use of Amazon Echo service, was implemented. The operation of the system is as follows: The patient activates the Amazon Echo device with his/her voice. The patient's input message uploads it to a cloud service, which is processed and converted, and then used by the custom application and decides if it will give feedback to the patient or will trigger interventions. Also, the application can offer dashboards of data to the patient.

A middleware service for the supervision of elderly health risk based on their activity and location at home is presented in [32]. The system provides home monitoring biometric and environment sensing data, via sensors, zone-aware location monitoring, context-aware activity monitoring, and decision-making treatment using the health risk ratio of the elderly. A cloud-based smart home server collects the data from sensors (bio-sensors and environmental), monitors the activity, locates the elderly's position, and analyzes and makes the decision based on various machine learning algorithms.

In project [33], a decision support cyber-physical system for assisted living and home monitoring is proposed. The system provides ambient monitoring and location detection of the supervised person and real-time alerting to the supervised person and his/her relatives. The system consists of two subsystems. The first subsystem monitors, via wireless sensors, the home's indoor conditions, for threshold violations. Then it communicates with a server, which in turn sends the real-time alerts, to the supervised person and his/her relatives. The second subsystem detects the monitored person's future locations, with the use of machine learning. The purpose of the implementation of this system is to assist older adults to maintain an autonomous lifestyle.

The design and implementation of a cyber-physical system for home monitoring for the elderly are presented in [34]. The system consists of an ambient tracking network of sensors and hardware devices and multi-platform software applications. The system's components provide advanced features like elderly's continuous indoor supervision, monitoring of home conditions, real-time alerts and notifications for the under-monitoring people and their caregivers, and interoperability with medical devices for tele-health purposes. The advantage of the system compared to other systems is the use of luminaire bulbs. These bulbs can be either smart or dummy and can provide information about indoor localization and home environment sensing, as well as receiving data from medical devices.

In [35] an IoT-based platform for smart homes is proposed. The aim of this platform is to provide elderly assistance to live longer in their home. This succeeds with the use of different technologies like radio-frequency identification (RFID), wearable electronics, wireless sensor networks, and AI. Thus, this platform is implemented with the use of sensors, actuators, and devices. These deviccs are wall lights for indoor localization, armchairs for sitting posture monitoring, belt for movement information, and wall panels and mobile devices as user interface.

A disappearing ubiquitous computing system to assist elderly people with dementia is proposed in [36]. The system exploits the acoustic stimulus to assist the sufferers of dementia to recognize their familiars. The innovation of this system is found on the non-intrusive way that integrates with the home and interacts with the elderly. With the use of smart devices, like smart speakers, smartphones, smartwatches, and smart-wristbands, the system identifies when a familiar person enters to the elderly's home, via the carried smart device. Then a sound, which has been correlated with that person, is reproduced via the smart speaker in order to stimulate the sufferer's memory.

It is noticed that the state of the art in home supporting smart systems for elderly people is moving toward human-centric systems. These systems are tailored to the needs of the human-user requirements and not vice-versa. Furthermore, another trend dictates adding intelligence to the home environment. This may be achieved with the exploitation of machine learning (ML), deep learning (DL), and artificial intelligent (AI) technologies and their integration to the support system. Also, technologies that add and evolve the intelligence of the home environment systems, while releasing the human-user from bearing many different devices onto him/her, are in need. In addition, the user physical interaction with the system is enhanced, turning the experience to interact with the system more natural. This eliminates the typical HCI, thus moving to a physical interaction approach. The benefits are the absence of need for special training of the user or home organization and furniture intervention that may upset the elders. Finally, it makes these solutions more appealing to older people since they adapt to their needs instead of the requirement for them to follow new rules in their households.

5 Technologies

As it was mentioned above, in order to implement a smart home, the use of different technologies is required. There are, however, three main ingredients that without them no home supporting systems may be implemented. These ingredients include the smart sensors, the communication equipment and protocols, and the storage that form the basic infrastructure for a smart home.

5.1 Sensors

A sensor is defined as any object that has the ability to detect and measure events and/or changes in the environment in which it is located. There are various types of capabilities of sensors, which can be integrated into a smart home infrastructure. In Fig. 1, a categorization of the sensors is presented as it was reported by Ahvar et al. [30]. Sensors are initially categorized into two main groups, static and mobile sensors. In the category of static sensors, sensors are embedded in the smart home

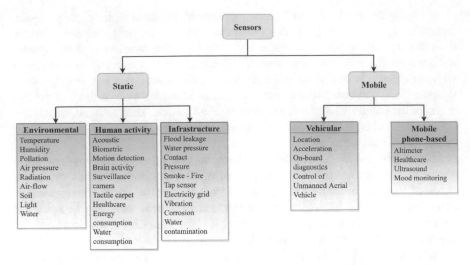

Fig. 1 The categorization of sensors

environment (e.g., thermostat), while, in the category of mobile sensors, sensors are usually used in wearable devices (e.g. smartwatch). At next, there is an additional categorization of the two main groups, based on the type of collected data from each type of sensor.

A further categorization of the sensors can be made based on their ability to communicate for the transfer of the collected data. There may be two categories. The smart sensors, which have the ability to communicate directly with the rest of smart home's infrastructure, while the second category of sensors requires the mediation of a platform to achieve the sensors' communication with the rest of the infrastructure.

5.2   Communication

The communication in a smart home is composed of two types of operations, as shown in Fig. 2. The first type is the communication of the objects within the house. While the second type of communication is related to the interconnection of the home with the outside world.

The communication within a smart home is divided into four different area networks.

- The BAN (body area network), which is the wireless network for wearable devices (or other devices on a user's body). They are mainly used for health-care purposes. Some of the technologies that apply to this type of network include NFC, RFID, and Ultra-WideBand (UWB – the wireless version of the USB).

Fig. 2 The categorization of communication in a smart home

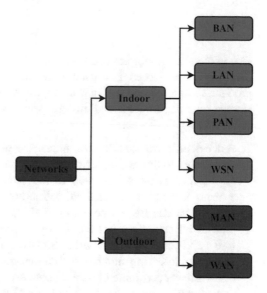

- The LAN (local area network) may be said to be the network that covers a range of approximately 1 km. Some of the technologies that are applicable to this type of network include Wi-Fi and Wi-Fi HaLow, ETSI HyperLAN, and WirelessHART.
- The PAN (personal area network). Some of the technologies that are applicable to this type of network include Bluetooth and Bluetooth Low Energy (BLE), ZigBee, Z-Wave, and 6LoWPAN.
- The WSN (wireless sensor network) is an alternative and most cost-effective way of interfacing sensors. They have quite low energy requirements compared to the other networks. Furthermore, WSN is the most common way of communicating for smart living purposes, as it is used in the surveillance and security systems, and for the energy management of a smart home.

In addition, the communication of the home with the outside world is also divided into two different area networks.

- The MAN (metropolitan area network) can transmit information within a radius of up to 20 km. Some of the technologies that are applicable to this type of network include WiMAX, ETSI HiperMAN, ZigBee-NAN (6LoWPAN), Wi-SUN (6LoWPAN), and NWave.
- The wide area network (WAN) is capable of transmitting the information over a wide geographical distance. Thus, the Internet may be considered as a WAN. In addition, the following are including in this category: 2G/3G/4G/5G, NarrowBand-IoT (NB-IoT), Low-Power Wide-Area Network (LPWAN), Wireless Regional Area Network (WRAN – IEEE 802.22), and Satellite communication and Radio cognitive [37–39].

5.3 Storage

One of the most important and valuable issue in smart homes is data. The collected data should be stored and used with an intelligent way toward the elderly's monitoring and supporting. In a smart infrastructure, such as a smart home, data storage can be implemented in two ways, either locally or as a cloud-based infrastructure.

1. A local infrastructure for the management and the storage of collected data contributes to the availability, reliability, security, and privacy of the data. This selection contributes generally to the security and safety of the whole home system. Examples of local infrastructure include Mobile Cloud Computing (MCC), Mobile Edge Computing (MEC), Cloudlet, and Fog Computing. This kind of infrastructure has the ability to process raw data, and reduce the data traffic and congestion that will eventually be transferred to the Internet.

2. The ever-increasing number of Internet-connected devices has as a result the ever-increasing volume of data that we are required to manage and store. Cloud technology seems to be the ideal solution to meet these requirements, since it offers low-cost, scalability options, user-friendly interfaces, a good degree of security, and, under certain circumstances, high level of availability and reliability. Thus, for managing and storing the collected big data, various Cloud-based platforms have been developed. These platforms may be categorized based on three models of cloud services: (i) Software-as-a-Service (SaaS), (ii) Infrastructure-as-a-Service (IaaS), and (iii) Platform-as-a-Service (PaaS). Some of these platforms include OpenIoT, Amazon, Google Cloud, Libellium, IBM Watson, FIWARE, Arkessa, OnePlatform, SensorCloud, SmartThings, ThinkWorx, Oracle IoT, Piotly, Nimbits, ThinkSpeak, Xively, etc. [39].

6 Future Trends

Many researchers have dealt with issues on designing and implementing various home supporting systems for elderly people. In addition, there are several other trends and challenges that future researchers need to face and propose solutions. In this section, future trends, hot topics, and challenges issues regarding the home supporting smart systems for elderly people are indicatively presented.

6.1 Robotics

Robots are a trend of automatic systems that are inserted in home environments to gather information, assist and interact with the elderly, and perform actions instead of the older people. The design, development, and extensive use of robots are quite promising for home assistance and support of the elderly. Robots can perform several tasks, like cooking, cleaning, etc., but also can serve as intelligent agents. Agents provide assistance to the elderly's daily life activities and act like companions, thus decreasing the degree of elderly's loneliness.

6.2 Machine Learning

Nowadays, machine learning (ML) is a very promising topic for research, attracting more and more researchers worldwide. It is a method used widely in many different disciplines and in a variety of services and applications. Regarding the scope of this chapter, ML is specifically popular in the smart home industry. With the employment of the ML to devices and systems of a smart home, personalized services (such as the recognition of daily-life activities and the supervision of the residents) may be achieved [40]. Therefore, ML method is imperative in the implementation for home support systems for the elderly, as through it the home will "learn" its residents. In this way, the constant support of the elderly in their daily routine can be achieved, as well as the intervention and decision-making from the home itself, in case of need.

6.3 Artificial Intelligence

Another current and future challenge is artificial intelligence (AI). AI is the way in which a computer system acquires intelligence. So, a smart home, and especially one, which is dedicated to supporting vulnerable groups of people like the elderly, is required to possess intelligence. Thus, it should be able to learn and adapt to the needs and demands of the elderly, to infer conclusions, and to understand their behavior and reaction, even solve problems. The integration of AI into physical environments created the topic of AmI [41]. Thus, a home supporting smart system should integrate AmI, allowing the continuous learning of human behavior and executing actions triggered by recognized events. This type of system can assist and support the elderly both in their activities and their routine, as well as emotionally.

7 Discussion

In this chapter, an overview of the research topic of home support systems for elders, exploiting smart technology was presented. The aim was to offer a survey for every researcher who wants to get involved in this topic and find in one place the state-of-the-art research results until now. Furthermore, we want to motivate researchers toward the topic of Silver Economy that urges for innovative solutions due to the population of the older people.

Focusing on smart home monitoring for the elders, the research results and the available commercial solutions were given. The overview revealed that few works have been presented by the academia and the industry toward this target group, although technology and the main components for designing a smart support home system for the elderly is quite mature. To facilitate a researcher discover easily the required knowledge to develop such systems, the chapter was divided into five main sections.

The first section (introduction is excluded) was dedicated to the historical evolution of the research topic, enabling a researcher to understand the requirements that led the research toward the proposed solutions. A thorough exploration of previous works is offered and detailed in order to depict the steps that were followed, mainly by academia, for achieving the development of home support systems. This section includes also commercially available solutions, highlighting that this topic is already mature to produce commercial products.

The second section is presenting the challenges that face the researchers nowadays. This is considered a critical part of the chapter, since it is significant for a researcher to understand the problems that need a solution. The most challenging issues in the implementation of IoT environments, such as the case of a smart home, are security and privacy. These two issues are presented in this section, and more details are given for the specifications and requirements that are associated to them.

In the next section, the state-of-the-art research is presented. Addressing partially the challenges that were mentioned in a previous section, researchers have proposed some solutions, which are considered to be the best at the moment of the writing of this chapter. The chapter presented a thorough overview of the state-of-the-art research on the topic, revealing that research tend to adopt solutions that do not affect significantly the day-to-day lives of the older people. The services that are identified by all works as important are those of position monitoring, fall detection, medication support, communication with the familiars and the caregivers, and services regarding well-being.

A very important section was the one that focused on the technologies that are used for developing the smart home system. In fact, three main ingredients include the smart sensors, the communication equipment and protocols, and the storage. A taxonomy of the sensors and the communication networks found in a smart home is given. The storage requirements are reported, and mature solutions are listed for use by the researcher.

The chapter closes with the report of three trends in research, which try to find their way to solve challenging issues. Namely, robotics, AI, and ML are the trends of high interest on the topic, with several solutions (of high cost) already available to the market. The most critical task of them all is to monitor the older person, without interfering to his/her life, and at the same time analyze the behavior as normal (expected) or not, and all of that without violating the resident's privacy, safety, and security.

It is the authors' belief that this overview of the literature and the market on this critical topic on Silver Economy will assist the future researchers to get involved more efficiently with the topic. The challenges for research are in some cases uncountable, but the potential for developing smart homes for the elders is unlimited, considering the number of the disciplines and the topics involved in this critical issue for our older fellows.

Acknowledgements The authors would like to thank Dr. Georgios Spathoulas for the appropriate guidance on security issues.

References

1. Zsarnoczky, M. (2016). Innovation challenges of the silver economy. *VADYBA, 28*(1), 105–109.
2. Jiang, L., Liu, D.-Y., & Yang, B. (2004). Smart home research. In *Proceedings of 2004 International Conference on Machine Learning and Cybernetics* (IEEE Cat. No. 04EX826, Vol. 2). IEEE.
3. IoT Technology Stack – from IoT Devices, Sensors, Actuators and Gateways to IoT Platforms. https://www.i-scoop.eu/internet-of-things-guide/iot-technology-stack-devices-gateways-platforms/. Cited 28 May 2019.
4. Aldrich, F. K. (2003). Smart homes: Past, present and future. In *Inside the smart home* (pp. 17–39). London: Springer.
5. Demiris, G., & Hensel, B. K. (2008). Technologies for an aging society: A systematic review of smart home applications. *Yearbook of Medical Informatics, 17*(1), 33–40.
6. SMARTHOME+. https://www.gosmarthomeplus.com/control/. Cited 28 May 2019.
7. D-Link Smart Home. https://eu.dlink.com/uk/en/for-home/smart-home. Cited 28 May 2019.
8. NETATMO. https://get.netatmo.com/renovation-en/. Cited 28 May 2019.
9. Homematic. https://www.eq-3.com/solutions/smart-home.html. Cited 28 May 2019.
10. iNELS. https://www.inels.com/apartment. Cited 28 May 2019.
11. Arcelus, A., et al. (2007). Integration of smart home technologies in a health monitoring system for the elderly. In *21st International Conference on Advanced Information Networking and Applications Workshops (AINAW'07)* (Vol. 2). IEEE.
12. Nourizadeh, S., et al. (2009). A distributed elderly healthcare system. In *Proceedings of the 1st International Workshop on Mobilizing Health Information to Support Healthcare-Related Knowledge Work, (MobiHealthInf)*.
13. Choukeir, A., et al. (2010). Health smart home. *International Journal of Computer Science Issues (IJCSI), 7*(6), 126.
14. Kim, J., et al. (2010). POSTECH's U-health smart home for elderly monitoring and support. In *2010 IEEE International Symposium on A World of Wireless, Mobile and Multimedia Networks (WoWMoM)*. IEEE.
15. Klack, L., et al. (2010). Future care floor: A sensitive floor for movement monitoring and fall detection in home environments. In *International Conference on Wireless Mobile Communication and Healthcare*. Berlin/Heidelberg: Springer.
16. Valero, Z., et al. (2010). AmIVital: Digital personal environment for health and well-being. In *International Conference on Wireless Mobile Communication and Healthcare*. Berlin/Heidelberg: Springer.
17. Suryadevara, N. K., et al. (2013). Forecasting the behavior of an elderly using wireless sensors data in a smart home. *Engineering Applications of Artificial Intelligence, 26*(10), 2641–2652.
18. Vuegen, L., et al. (2013). Automatic monitoring of activities of daily living based on real-life acoustic sensor data: A preliminary study. In *Proceedings of the Fourth Workshop on Speech and Language Processing for Assistive Technologies*.
19. Panagopoulos, C., et al. (2015). Evaluation of a mobile home care platform. In *European Conference on Ambient Intelligence*. Cham: Springer.

20. Dasios, A., et al. (2015). Hands-on experiences in deploying cost-effective ambient-assisted living systems. *Sensors, 15*(6), 14487–14512.
21. Korres, S. P., et al. (2018). A low-cost IoT-based health monitoring platform enriched with social networking facilities. In *2018 IEEE International Conference on Pervasive Computing and Communications Workshops (PerCom Workshops)*.
22. Bassoli, M., Bianchi, V., & Munari, I. (2018). A plug and play IoT Wi-Fi smart home system for human monitoring. *Electronics, 7*(9), 200
23. Yang, D., et al. (2018). Passive infrared (PIR)-based indoor position tracking for smart homes using accessibility maps and a-star algorithm. *Sensors, 18*(2), 332.
24. Voicu, R.-A., et al. (2019). Human physical activity recognition using smartphone sensors. *Sensors, 19*(3), 458.
25. Panwar, N., et al. (2019). Smart home survey on security and privacy. arXiv preprint arXiv:1904.05476.
26. Pirbhulal, S., et al. (2017). A novel secure IoT-based smart home automation system using a wireless sensor network. *Sensors, 17*(1), 69.
27. Maglogiannis, I. (2014). Human centered computing for the development of assistive environments: The STHENOS project. In *Proceedings of the 7th International Conference on PErvasive Technologies Related to Assistive Environments*. ACM.
28. Maglogiannis, I., et al. (2014). Mobile reminder system for furthering patient adherence utilizing commodity smartwatch and Android devices. In *2014 4th International Conference on Wireless Mobile Communication and Healthcare-Transforming Healthcare Through Innovations in Mobile and Wireless Technologies (MOBIHEALTH)*. IEEE.
29. Kakarountas, A. (2014). Disappearing computing for elderly assisted living. In *2014 4th International Conference on Wireless Mobile Communication and Healthcare-Transforming Healthcare Through Innovations in Mobile and Wireless Technologies (MOBIHEALTH)*. IEEE.
30. Ahvar, E., et al. (2016). Sensor network-based and user-friendly user location discovery for future smart homes. *Sensors, 16*(7), 969.
31. Bafhtiar, G., et al. (2017). Providing patient home clinical decision support using off-the-shelf cloud-based smart voice recognition. In *WIN 2017 Conference CDS Stream*.
32. Jung, Y. (2017). Hybrid-aware model for senior wellness service in smart home. *Sensors, 17*(5), 1182.
33. Bocicor, M. I., et al. (2018). Intelligent decision support for pervasive home monitoring and assisted living. In *2018 IEEE 14th International Conference on Intelligent Computer Communication and Processing (ICCP)*. IEEE.
34. Marin, I., et al. (2018). i-light-intelligent luminaire based platform for home monitoring and assisted living. *Electronics, 7*(10), 220.
35. Borelli, E., et al. (2019). HABITAT: An IoT solution for independent elderly. *Sensors, 19*(5), 1258.
36. Borelli, E., et al. (2019). An acoustic-based smart home system for people suffering from dementia. *Technologies, 7*(1), 29.
37. Ricquebourg, V., et al. (2006). The smart home concept: Our immediate future. In *2006 1st IEEE International Conference on e-Learning in Industrial Electronics*. IEEE.
38. Mendes, T., et al. (2015). Smart home communication technologies and applications: Wireless protocol assessment for home area network resources. *Energies, 8*(7), 7279–7311.
39. Čolaković, A., & Hadžialić, M. (2018). Internet of things (IoT): A review of enabling technologies, challenges, and open research issues. *Computer Networks, 144*, 17–39.
40. Cook, D. J. (2010). Learning setting-generalized activity models for smart spaces. *IEEE Intelligent Systems, 2010*(99), 1.
41. Ramos, C., Augusto, J. C., & Shapiro, D. (2008). Ambient intelligence – The next step for artificial intelligence. *IEEE Intelligent Systems, 23*(2), 15–18.

Human–Computer Interaction Technique for Irrigation and Sun Tracking Solar Panel Model

F. Pradeep, Neena Boneyfus, Shilpa Theres, S. Gokul, and Sagaya Aurelia

1 Introduction

Agriculture is defined as the premise for life for the humans as it's the principle wellspring of raw materials and other food grains. It plays an important role in the development of the economy of the country. As occupation is the major crisis, through agriculture, it provides a greater opportunity to many people. Development in agriculture is essential for the improvement of the financial state of the nation. Shockingly, numerous farmers still utilize the conventional techniques for cultivating due to which the results in yield are low. Agriculture is said to be the basic of life for humans since it is the one and only important source of raw materials and other food grains. It is an important part in the development of a nation's economy. It additionally gives huge work chances to the general population who are in need for the job. Development in farming part is essential for the advancement of financial state of the nation. Replacing humans by automated machines has definitely improved the yield. Therefore, there has to be a technology that needs to be implemented in the agriculture division to expand the yield of the harvest. The utilization of wireless sensor is arranged so that it gathers data from various kinds of sensors and therefore sends it to the server utilizing wireless protocol. The data from the environmental factors are collected, and this helps in monitoring the system.

The irrigation scenario is divided due to greater demand for a better productivity, poor performance, and because the water that is used in agriculture is reduced. These problems can be solved by using an automated system that can be used in irrigation.

F. Pradeep (✉) · N. Boneyfus · S. Theres · S. Gokul · S. Aurelia
Department of Computer Science, Christ University, Bengaluru, India
e-mail: pradeep.f@cs.christuniversity.in; neena.arakkal@cs.christuniversity.in;
shilpa.theres@cs.christuniversity.in; gokul.s@cs.christuniversity.in;
sagaya.aurelia@christuniversity.in

© Springer Nature Switzerland AG 2020
S. Paiva, S. Paul (eds.), *Convergence of ICT and Smart Devices for Emerging Applications*, EAI/Springer Innovations in Communication and Computing,
https://doi.org/10.1007/978-3-030-41368-2_5

Automatic irrigation is required because

- It is easy and very simple to configure and also in installing.
- It also saves a huge amount of resources and energy so it can be used efficiently.
- Farmers also will maintain correct amount of water at the right time if the irrigation is automatic.
- If irrigation is avoided from the wrong time of day, it will decrease the runoff from overwatering the saturated soil that will automatically improve the performance of the crops.
- This automated system uses valves that are used for the turning ON and OFF of the motor. Therefore, there is no labor required for the turning ON and OFF of the motor.
- There is also a manual system that is developed in which the user according to the requirement of water of the crop, he himself can ON and OFF the motor.
- This system is very valuable for controlling the moisture content in the soil in a highly specialized greenhouse vegetable production.
- It also saves time and the chances for human errors are also eliminated.
- This system helps in supporting the water management system; it uses a GSM module for monitoring the whole system.
- The system will monitor the water content in tank and provide accurate quantity of water as per the requirement of the crop.
- The system will keep track of the temperature and humidity of the soil which help to maintain the composition of nutrients that is used for the growth of the crop.
- It is very effective and of lower cost with less consumption of power by using sensors that are used for controlling as well as monitoring the devices that can be controlled by SMS by using a GSM (Android mobile).
- The sun tracking solar panel is used to save energy. (LDRs move in the direction of the sun, and the energy is stored within the batteries, and this can be used in the motor pump that uses electricity to pump water to the crops.)
- This system conserves electricity and also water.

1.1   Irrigation Using IoT

Checking only with the factors of the environment is not sufficient, and it is not enough to improve yield of the harvests. The productivity is influenced to a greater extent due to different components. These components can be bugs and insects or pests, which can be constrained by showering the harvest with bug spray and pesticides, also due to birds and animals at the point when the yield grows up. There is additionally a probability of theft at the point when yield is at the phase of reaping. Indeed, even after harvesting, farmers additionally face issues of harvest that has been collected and stored. Along these factors, so as to give answers for the issues that are being faced, it's important to create an integrated system that will deal with all variables influencing the production in each stage like, development, gathering, and, furthermore, postharvest and its storage.

Therefore, this chapter provides a system which is useful in controlling several operations that have to be performed in the field that give flexibility and also in maintaining and monitoring the field data. The main motive of this chapter is to make agriculture smart by making use of IoT techniques as well as automation.

The main motive of an IOT smart irrigation model is to make the work of the farmers easier. There are several sensors that are used. The sensors are used to sense the soil moisture, temperature, and humidity, and if the moisture content is very low, then the motor is automatically switched on and will be turned off after getting the required water to the soil. The notifications are also sent based on the respective tasks. For this, there is a GSM that is used to send the notification to the farmers.

Figures 1 and 2 depict how different technologies are being used in smart irrigation.

1.2 Sun Tracking Solar Panel

Solar energy is ending up progressively rewarding with the expanding cost and consistent exhaustion of nonrenewable sources and the developing interest of the other renewable power sources, for example, solar wind, geothermal, and tidal wave.

Fig. 1 IoT in agriculture

Fig. 2 Technology used in agriculture

Be that as it may, despite the various advantages of solar energy, solar boards which catch light are stationary (solar exhibit has a fixed direction to the sky). These stationary just-as-costly solar boards will not be able to extract a good amount of solar energy as the weather conditions are not stable.

The power outcome of solar panel is most extreme when it is situated oppositely to the bearing of sun beams as both the zone of brightening of daylight on solar panels and force of sun beams are greatest for this situation. It has been discovered that the proficiency of solar panels improves by 30–60% when we utilize a portable solar following framework rather than a stationary cluster of solar panel.

The plan and execution of a power productive solar tracker are along these lines a test attributable to the stability of the solar panels. The edge of tendency of sun beams with the solar panels changes continuously because of the development of the sun from east to west in light of the world's rotation. Also, during the days that are cloudy, the circumstance absolutely gets wild. Moreover, the Earth's revolution adjusts the separation among the Earth and the sun, which presents change of, for example, approaching sun beams. Every one of these elements ought to be remembered for structuring the solar tracking for electricity to accomplish greatest productivity (Fig. 3).

This chapter discusses about the solar tracking system that has been designed using four LDRs and two servo motors. The main aim of this model is to sense the intensity of the sun rays that is sensed by LDRs. There is also a microcontroller by

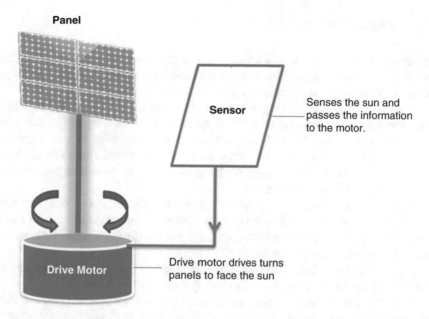

Panel

Sensor

Senses the sun and passes the information to the motor.

Drive Motor

Drive motor drives turns panels to face the sun

Fig. 3 Introduction to solar panel

which the motor will be rotated based on the direction of the light that is falling on the LDR. The intensity of the sunlight will change depending on the weather conditions and also the position of the device, for which there is a way by which the threshold values can be changed by using variable resistance.

The system has been proposed to support water management in agriculture. Since there is scarcity of water everywhere, it has to be used efficiently. The sun tracking solar panel will also help to save electricity, and thus it will reduce the power consumption too. It can be used in areas which need proper water management in irrigation. This system will have a huge demand in the future. This mainly reduces the human errors in adjusting the moisture in the soil levels, and it also helps to maximize the net profits according to the quality, sales, as well as the future growth of the product.

In this chapter Sect. 2 discusses the literature review; Sect. 3 has the pictorial block diagram of the whole smart irrigation system; Sect. 4 discusses about the various components required to build the system; Sect. 5 discusses about how the components are connected to the Arduino, and the circuit diagram of the same is also represented; Sect. 6 explains the working of the smart irrigation and sun tracking solar panel model; Sect. 7 discusses how the project is tested and evaluated; Sect. 8 depicts the results obtained from the system; Sect. 9 discusses about the limitations of this project; and Sect. 10 elaborates the future scope and conclusion of the project.

2 Literature Review

In paper [1] the authors have discussed about the smart irrigation project which has moisture and temperature sensors which have en-cooperated both Arduino and breadboard connections to make this model work.

In paper [2] the authors have discussed about the smart irrigation using GSM. Here the automated system is designed to make the tank refill system so that water level can be checked in the tank and then water can be used for irrigation process. In paper [3] the author discusses about the sound and alarm system in the home automation system; from this paper we learn about how GSM can be implemented and how to implement in this model.

In paper [4] the author uses the Arduino-based smart irrigation to check the water level, to check the water level of the sink, rainfall automation system and this can be used for the smart irrigation project so that the yield of the farmers can be increased and the conservation water can be more.

In paper [5] the author discusses about the smart irrigation project and how the water pump can be used for manually turning the water pump on/off.

In paper [6] the author discusses about smart irrigation project and how the water can be drip irrigated using the GPRS and GSM modules by connecting it to Arduino board.

From this website [7], they discusses about implementing sun tracking solar panel using four LDRs; how this model can be implemented and other uses by this model are also being discussed.

In paper [8] in this paper, the author discusses about the sun tracking solar panel, but in this model, the author has used two LDRs and one servo motor so that the sun cannot track in all directions; it will be difficult to conserve more energy from it.

In paper [9] the authors have discussed the importance of solar energy and how this can be conserved and used for various applications. In this case they have used four LDRs and servo motors so that the solar panel can track the sunlight in all directions; this model is being developed using Arduino and breadboard connections in it.

In website [10], the website discusses about how the smart irrigation has improved the farmer's life and how this smart irrigation is more beneficial for farmers.

In website [11], this site discusses about the importance of smart irrigation and how it is making farmers' life easy and how this can be used to increase the profit of the farmers. In website [12], they discuss about the sky sprinkler methodology for irrigation system so that the water conservation can be made more efficient; this is one of the best technologies so that it can be water can be saved and there is no wastage of water; by using this technology there are various technologies. In paper [13] the authors have discussed about an array of sensors, that is, humidity sensors, methane sensors, and various other sensors which can help the farmers to get approximate result regarding the environment. In paper [14] the authors have discussed about the various ways the wireless technology can be used for smart irrigation project and also have discussed about having a wireless communication system in any model which is being developed. In paper [15] the authors have discussed about how the solar energy can be used for multiple uses. In this model, the author has saved the solar energy and used that to pump water from the bore well and other outlets. So by this model, the water can be conserved and used efficiently.

In paper [16] the author discusses about the working of the solar sun tracking panel using Arduino and the different servo motors. In the model the solar panel which is obtained is being preserved and used for future use. In paper [17] the author has explained about the ANSI programming language, which is very important for the Arduino programming. In paper [18] the author discusses how the microcontroller can be used for sun tracking, and the model is being demonstrated and tested. In paper [19] the author discusses about the wireless controlling of drip irrigation system; uses of drip irrigation system are also being mentioned in the paper. In paper [20] the author discusses about how drip irrigation can be used to save water and how this can be modeled with the use of Arduino so that wireless connection among them can be made.

3   Block Diagram of the Model

Figure 4 depicts a block diagram of the whole model which is implemented.

4   Components Required

Figure 5 illustrates the components that are required to develop this whole smart irrigation and solar tracking panel.

4.1   Arduino Board

An Arduino as shown in the Fig. 6 is an open hardware which is used to build devices that interact with the real world. An Arduino board uses various controllers and microcontrollers. These are basically designed with sets of analog and digital input and output pins that can be used for developing any devices and can also be interfaced to other circuits and boards. It can interact with LEDs, GPS units, television, smartphones, etc.

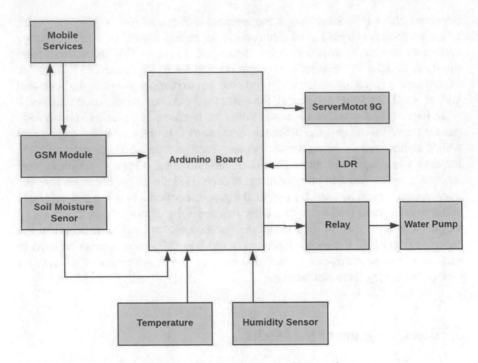

Fig. 4 Block diagram of smart irrigation and solar sun tracking panel

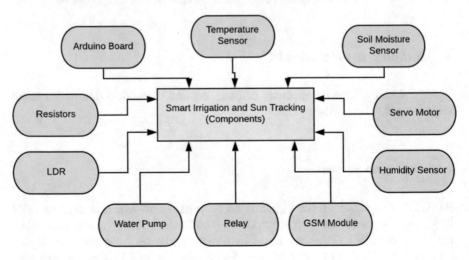

Fig. 5 Components required for model

Fig. 6 Arduino board

Fig. 7 Soil moisture sensor

4.2 Soil Moisture Sensor

The soil moisture sensor as depicted in Fig. 7 uses capacitance to check the volumetric water content of the soil. Just by inserting the rugged sensor into the soil which is to be tested, the result of the same is being displayed as the percentage of the water content in the soil. A dielectric permittivity function is used to measure the water content in the soil. The dielectric constant can be thought of as the soil's ability to transmit electricity. As the water content of the soil increases gradually, the dielectric constant of the soil also increases accordingly.

4.3 Temperature Sensor

Temperature is the most regularly checked or measured parameter in the world. This sensor as shown in Fig. 8 is used in our households items like ovens, refrigerators,

Fig. 8 Temperature sensor

and air conditioners. This is mostly used in all fields of engineering works. It measures the heat and cold (temperature) of the object to which the temperature sensor is connected. It generates a current or voltage output which is then measured or processed as per the application is designed.

Here in this model, we use the LM35 sensor to measure the temperature. This LM35 produces a linear output with 10 mV/degree scale, and these outputs can be given to different circuits, so after these outputs are being given, it can alert according to the inputs which are being taken, or it can be also connected to a relay to controller the temperature accordingly.

4.4 Humidity Sensor

Humidity can be defined as the amount of water content present in the air, according to the climate. This sensor as shown in Fig. 9 is used for measuring and monitoring humidity. These sensors are most widely used in different applications like industrial, biomedical, environment, and etc.

4.5 Relay

Relay as seen in Fig. 10 is a switch that is electrically operated. The main advantage of using the relay is that it takes a small amount of current to operate the relay coil. These relay can be used to control heaters, lamps, circuits, or any motors. There are various versions of relay which are available; they have their own applications in the circuits which they are used in. Basically these relays can be operated in two ways: firstly, it is used as an electromagnet to mechanically operate the switch and secondly, it can be used as a solid state relay.

Fig. 9 Humidity sensor

Fig. 10 Relay

Fig. 11 LDR

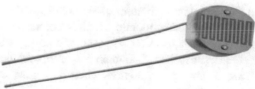

4.6 LDR

The main impulsion is to design a high-quality solar tracker. This basically works on the principle of photoconductivity. The resistance value decreases as the intensity of the light decreases. The basic passive component is the resistor in the LDR. It can simply be defined as the photosensor. These are very useful in dark and light sensor applications or circuits. This LDR measures the physical quantity of the light and then converts it into a signal it can read by the observer, or it can be sent to other circuits which operate further on it. It has a simple structure and low cost too. Figure 11 shows how the LDR looks.

Fig. 12 Servo motor 9g

4.7 Servo Motor 9g

This servo motor as shown in the Fig. 12 are small and lightweight motors, which are controlled with help of the servomechanism. A single servo motor can rotate 180° approximately. It can program using servo codes or hardware to control these servo motors. It uses servomechanism, and also it uses DC motor mechanism for controlling the rotation of the motor in a certain angular position. The main use of the servo motor is that it has angular precision, that is, it can stop at the certain position and can wait in the same position until the next signal is being obtained.

4.8 GSM Module

It is a cellular communication module which communicates between the GSM module and to the mobile which is connected too. It stands for global system for mobile communication. This module consists of base station, mobile station, and network subsystem. The main use of this module is for sending messages from the GSM to the mobile station. There is a SIM which is being attached within this module, so this acts as the main source for the communication between the two entities (Fig. 13).

5 Circuit Connection

5.1 Smart Irrigation System

The whole project is controlled and developed using the Arduino. The soil moisture is used to check the moisture of the soil, the output of the soil moisture sensor is connected to the digital pin D7 of the Arduino, and the humidity and the temperature

Fig. 13 GSM module

sensors are being connected to the digital pin of D2 and D3 of the Arduino, respectively. It uses an LED at the soil sensor circuit; this indicates that if the LED's on, then there is absence of soil moisture; if the LED's off then there is presence of soil moisture in the soil.

To communicate the output from the Arduino to the farmer, we use a GSM module. This is used for sending messages from the Arduino board. In developing this smart irrigation model, we use SIM800, which can give and take TTL logic directly. This also uses a voltage regulator, to power the SIM800 GSM module.

User can use two types of GSM module. It can use either use SIM900 TTL module, then 5 V power supply must be used, or it can use SIM900 module, then it applies 12 V power in the DC jack slot of the Arduino board.

12 V relay is used to control water pump which is deployed in the smart irrigation model. It is driven by a BC547 transistor which is again connected to the Pin 11 on the Arduino board. Figure 14 represents the circuit diagram of the smart irrigation project.

5.2   Sun Tracking Solar Panel

This model use 4 LDRs and two servo motors. Each LDR is connected to 100 kΩ resistors, and voltage divider fashion is used for the connection between LDRs. The outputs are being given to four analog inputs of the Arduino. Figure 15 has only two LDRs in it; in this project we use four LDRs so the same circuit diagram can be referred for the same.

After the inputs are being received from the LDRs, it's processed; these outputs are being given as the PWM inputs of the two servo motors from the digital pins 9 and 10 of the Arduino.

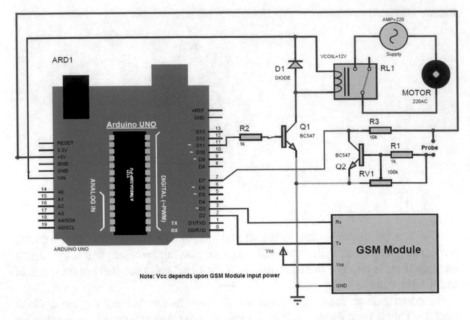

Fig. 14 Circuit diagram of smart irrigation model

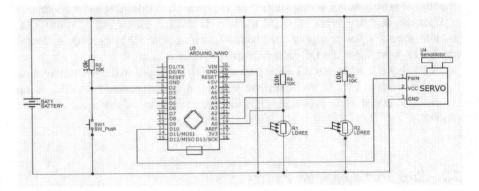

Fig. 15 Circuit diagram of sun tracking solar panel

6 Working

6.1 Smart Irrigation

Arduino is connected to different sensors like humidity sensor, temperature sensor, and soil moisture sensor. These sensors are used to check the soil moisture content of the soil, the humidity of the air, and the temperature of the environment, respectively. The outputs obtained from each sensor are being sent to the Arduino and are being analyzed accordingly. The GSM module that is connected to the Arduino is used to send a message to the mobile phone about the data that is collected. The GSM module can also be used to manually operate the water pump. After the analysis the GSM module is used to send the information to the mobile phone.

The soil moisture sensor, humidity sensor, and temperature sensor are connected to the analog pins of the Arduino board as showed in the circuit diagram. The USB cable generates 5 V that is connected to the Arduino. First of all the GSM module is initialized, and it waits until it connected to the network. The data that is sensed from the sensors are sent to the microcontroller. There is threshold value already set, so if the moisture content in the soil is below the threshold value, then automatically the water pump is on. The water pump can also be controlled by sending messages to the GSM module.

6.2 Sun Tracking Solar Panel

As the solar panel is fixed, it decreases the efficiency. The efficiency can only be increased when the intensity of light is perpendicular to the solar panel. In order to increase the efficiency of the solar panel, we use LDR that is used to detect the sunlight and also motors for the movement of the solar panel according to the direction of sunlight. The orientation of the solar panel is determined by the changes in LDRs when the sunlight falls on them. LDRs are used as the main light sensors. Two servo motors are fixed to the structure that holds the solar panel. The program for Arduino is uploaded to the microcontroller. LDRs sense the amount of sunlight falling on them. Four LDRs are divided into top, bottom, left, and right.

For east–west tracking, the analog values from the two top LDRs and two LDRs at the bottom are taken and are analyzed. The solar panel would move in the direction where more light is falling on the LDR. If the set of LRDs that is connected to the bottom receive more light, then the solar panel will move in that direction. The left set of LDR is used for angular deflection. When the left set of LDR receives more light compared to the right set, then the horizontal servo motor will start moving, and if right set of LDR receives more light compared to left set of LDR, then the servo motor will move in that direction.

7 Testing

The model was developed with all the components which were listed above. The same circuit model as shown in Fig. 14 above was developed and as shown in Fig. 16. The model developed was tested with all constrains that are related to the model. Whenever the soil moisture is low, a message is sent to the number which is being deployed in the GSM module, so accordingly the water pump is in on state. As soon as the soil moisture reaches 100%, the water pump is in the off state after

Fig. 16 Implementation model of smart irrigation

5 seconds. To control the water pump manually, we use Bluetooth to control the water pump from the location where the user is located. But this works only when the user is 200–250 meters far.

The output screen after testing is being shown in Fig. 17 below. Here AT+CMGF represents that the GSM module is ready to send the message. It is a binary value as it either takes it as 1 or 0. AT+CMGS has the 10-digit number to which the message has to be sent. These numbers are being predefined in the program.

As shown in Fig. 18, sun tracking solar panel model is also being developed. This model was tested with the torch light. It is working as it was described in the working module; it rotates approximately 360° to track the sun rays.

8   Results and Discussion

This project was mainly designed to overcome the difficulties faced by the farmers. The result obtained from the model is that whenever the soil moisture, humidity sensors, temperature sensors output of the sensors are being sent to the Arduino. The result is being verified according to the given range, and then the output obtained is being sent to the GSM module; message or SMS is being sent to the farmer's mobile for every instance of time. Then if the moisture in the soil is low, then the water will be automatically pumped. Once the soil moisture is obtained, then the pump is turned off. Accordingly different plants have natures; some plants takes more water and some take less water, so by taking this into consideration, the water on/off can be also handled from the mobile itself.

The sun tracking solar panel is also added as a part to this smart irrigation project. The designed tracker works successfully in all the directions. The sun tracking solar panel can track the rays of the sun, meaning the sun rays, by using microcontroller

Fig. 17 Output screen of smart irrigation project

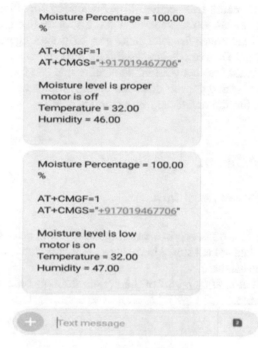

Moisture Percentage = 100.00 %

AT+CMGF=1
AT+CMGS="+917019467706"

Moisture level is proper
 motor is off
Temperature = 32.00
Humidity = 46.00

Moisture Percentage = 100.00 %

AT+CMGF=1
AT+CMGS="+917019467706"

Moisture level is low
 motor is on
Temperature = 32.00
Humidity = 47.00

Text message

Fig. 18 Sun tracking solar panel model

and servo motor. Irrespective of the fluctuating conditions in weather and locations, this system can work properly. It can also initialize the starting position once the sun sets. Moreover, the solar panel goes to its starting position during the night. The number of LDRs can be increased for more accurate output.

The basic idea between implementing this sun tracking solar panel is to absorb the sun rays and convert this solar energy into electrical energy and then it can be used for turning on/off the water pump or so on.

9 Limitations

9.1 Smart Irrigation

1. If there is no network, then using GSM would be of no use.
2. The data are not stored in it, that is, output which is obtained is not getting stored in the database.
3. The automating on/off of the system through Bluetooth is only for a minimum distance.

9.2 Sun Tracking Solar Panel

1. When the atmosphere is cloudy, it will be difficult to track the sun.
2. Placing the LDR in a perpendicular degree to sunlight is troublesome.
3. LDRs are delicate elements and therefore can get harmed during severe climatic conditions.
4. We can save the solar energy into batteries, but they are heavy and will occupy more space. It will be troublesome to change it from time to time. Solar panels are usually expensive.
5. Micro servo motor should be replaced by other motors for external application for large energy production.

10 Future Scope and Conclusion

There is always a scope for improvement. In the future we will have more new technologies coming up which we can implement. Right now the things we can think of for future improvement are

1. Replacing the GSM with Wi-Fi module for better working.
2. Introducing an app for the same; therefore, the work can be done quickly and easily.

3. From the sun tracking solar panel, we can trap the sun's rays and convert it into electrical energy and that can be preserved and can be used for other works.

The world is growing into new advances, and it is necessary to trend up in agriculture. Numerous researches are done in the field of agriculture. The gathered information gives the data about the different environmental factors. Observing the ecological elements is not the complete answer for incrementing the yield of crops. There are a number of other factors that decrease the productivity to a greater extent. The IoT agricultural applications are making it possible for farmers to collect significant data. Small farmers must know the potential of IoT for agriculture by implementing smart technologies to grow the competitiveness and sustainability in the making. With the population growing quickly, the demand can be successfully met if the small farmers execute agricultural IoT solutions in a good manner. Hence automation can be implemented in agricultural sectors to overcome problems.

References

1. Nayyar, A., & Puri, E. V. (2016, November). Smart farming: IoT based smart sensors agriculture stick for live temperature and moisture monitoring using Arduino cloud computing & solar technology. In *Conference: The international conference on Communication and Computing Systems (ICCCS-2016)*.
2. Saha, H. N., Banerjee, T., Saha, S. K., Das, A., Dutta, A., Roy, A., & Chakravorty, N. (2018). Smart irrigation system using Arduino and GSM module. In *2018 IEEE 9th annual Information Technology, Electronics and Mobile Communication Conference (IEMCON)*. https://doi.org/10.1109/iemcon.2018.8614839.
3. Ying-Nan, L., et al. (2017). Laser alarm system based on GSM module family. *International Journal of Smart Home, 11*(5), 41–50. https://doi.org/10.14257/ijsh.2017.11.5.04.
4. Mahesh, V., Subba Rao, D., & Subbanna, S. (2015). Irrigation system using a wireless sensor network and GPRS. *International Journal of Innovation Technologies, IEEE Ago, 3*(7), 1154–1160.
5. Tiwari, M., Shukla, S., Sharma, V., Sahu, M., Sinha, H., & Singh, D. (2018). GSM-based automated and smart irrigation system. *i-Manager's Journal on Instrumentation and Control Engineering, 6*(1), 25–30. https://doi.org/10.26634/jic.6.1.13936.
6. Thakali, S. GSM based automatic irrigation system. Academia.edu. www.academia.edu/9587271/GSM_based_Automatic_Irrigation_System
7. GSM modem interfacing with Arduino. Circuits4you.Com, 22 July 2018. circuits4you.com/2016/06/15/gsm-modem-interfacing-arduino/
8. Arduino based sun tracking solar panel project using LDR and servo motor. circuitdigest.com/microcontroller-projects/arduino-solar-panel-tracker
9. Solar Tracking System – IJSER. www.ijser.org/researchpaper/SOLAR-TRACKING-SYSTEM.pdf
10. A low cost smart irrigation control system – IEEE Conference Publication. ieeexplore.ieee.org/abstract/document/7124763
11. What is smart irrigation? HydroPoint, 14 May 2019. www.hydropoint.com/what-is-smart-irrigation/
12. Skydrop – smart sprinkler controller. Skydrop smart sprinkler controller. www.skydrop.com/smart-irrigation-systems-a-greener-idea/.

13. A real-time wireless smart sensor array for scheduling irrigation. *Computers and Electronics in Agriculture*. Elsevier, 4 Sept. 2007. www.sciencedirect.com/science/article/pii/S0168169907001706
14. US5333785A – Wireless Irrigation System. Google Patents, Google. patents.google.com/patent/US5333785A/en
15. Harishankar, S., Sathish Kumar, R., Sudharsan, K. P., Vignesh, U., & Viveknath, T. (2014). Solar powered smart irrigation system. *Advance in Electronic and Electric Engineering*. ISSN 2231-1297, *4*(4), 341–346.
16. Ramya, P., & Ananth, R. (2016, August). The Implementation of solar tracker using Arduino with servomotors. *International Research Journal of Engineering and Technology (IRJET)*, *3*(8), 2395-56.
17. Balagurusamy, E. (2008). *Programming in ANSI C*. New Delhi: Tata McGraw-Hill Publishing Company Limited.
18. Filfil, N. A., Mohussen, D. H., & Zidan, K. A. (2011). Microcontroller-based sun path tracking system. *Engineering and Technology Journal, 29*(7), 1345–1359.
19. Dursun, M., & Ozden, S. (2011, April 4). A wireless application of drip irrigation automation supported by soil moisture sensors. *Scientific Research and Essays, 6*(7), 1573–1582.
20. Singh, S., & Sharma, N. (2012, August). Research paper on drip irrigation management using wireless sensor. *International Journal of Computer Networks and Wireless Communications*, *2*(4), 461–464.

HCI Authentication to Prevent Internal Threats in Cloud Computing

Hyeon Choe and Sagaya Aurelia

1 Introduction

1.1 Background

Recently, due to various and rapid developments in IT technology and development of highly specialized devices, information sharing and data use are freely performed anytime and anywhere. Currently, cloud computing services that provide virtualized data can be seen as key technologies that represent these changes in IT technology.

A cloud computing service is a type of computing service that stores large data resources in a virtual space, serves data resources with distributed processing technology, and pays for the services received [1]. These services are time and space independent and have the advantage of being easy to use even without special infrastructure expertise or IT skills. However, these factors can be vulnerabilities that can cause big problems if viewed from a security point of view.

Currently, solution vendor companies developing and delivering cloud services are mobilizing staff and releasing new security service offerings for billions of dollars in annual costs and maintenance due to security issues. Importantly, the focus is on strengthening security for virtual machines and networks. However, in addition to outside intrusion, the threat of information taking by insiders is also serious. In order to fully protect personal information and company secrets, it is urgent to prepare defenses against insiders.

H. Choe (✉) · S. Aurelia
Department of Computer Science, CHRIST (Deemed to be University), Bengaluru, India
e-mail: choe.hyeon@mca.christuniversity.in; sagaya.aurelia@christuniversity.in

© Springer Nature Switzerland AG 2020   119
S. Paiva, S. Paul (eds.), *Convergence of ICT and Smart Devices for Emerging Applications*, EAI/Springer Innovations in Communication and Computing,
https://doi.org/10.1007/978-3-030-41368-2_6

1.2 Necessity and Purpose

The malicious insider of cloud computing can exist for both cloud service providers and customers using cloud computing. If the threat is judged based on the accessibility of the information, the malicious insider of the provider will be a greater threat. Technical reinforcement of manager certification is required to play the role of provider to protect vast data.

HCI technology should basically provide convenience to the user and allow easy access to the computer. With the development of technology, many new HCI technologies have been born, which have been developed not only for convenience but also for strengthening user authentication.

Unlike the existing computing services, cloud computing services have a huge amount of personal information and data resources. Therefore, it is time for a security authentication system service suitable for a cloud computing environment rather than supplementing existing service security functions. In addition, security management systems, such as data protection, availability, and privacy protection, are important because they are applied in various service areas due to the development of cloud computing technology and various implementations [2].

The steps of the security management system must be realized from personal authentication. Due to the characteristics of cloud environment, it is necessary to strengthen authentication of providers and administrators who need more security. The most effective method for personal authentication will be the multiple utilization of HCI technology. The use of this enhanced personal authentication method is expected to lead to greater trust between vendors and users of cloud computing services and further expansion of cloud computing services.

2 Malicious Insider Threats

Cloud services are a form of outsourcing some or all without ownership of IT resources, so security issues are inevitable. The security concerns of using cloud services include security and data leakage, service reliability and availability, interworking with existing application services, stability of service providers, compliance, and service costs.

Table 1 shows which threats are important issues in cloud computing. Recent examples show that information leaks by malicious insiders are extremely fatal [3]. In 2017, a US carrier, Verizon, reported a security breach that leaked 14 million customer data. Key information from tens of millions of customers was located on Amazon's public cloud servers; it was reported that access to the entire data was available for download because there was no control over access rights [4]. The public cloud service was Shadow IT, which was not checked or managed by IT staff and is known as an insider threat. Similarly, insider threats involving the Republican National Committee in 2017 resulted in the disclosure of 198 million

Table 1 Top threats to cloud computing

Classification	Explanation
Abuse and nefarious use	When people with malicious intentions use the cloud, it can be harder and more difficult to find than traditional botnets because all information is virtual.
Insecure interfaces and APIs	If someone reuses or synthesizes existing code to expedite application building, there will be weak spot in security.
Malicious insider	If we hire people quickly, we are more likely to hire people who are not morally fit.
Shared technology issues	Failure to properly manage a virtual machine can threaten the entire system with a single small mistake.
Data loss or leakage	Existing controls to protect data may not be suitable in a new cloud environment and are harder to monitor.
Account or service hijacking	It is vulnerable to many hijacking techniques already used to redirect to various malicious sites.
Unknown risk profile	Often, service providers are less transparent, and customers do not know if they need to run a system configuration or a software patch.
Distributed denial of services	There is an increasing number of ways to attack specific sites by distributing and deploying multiple attackers simultaneously.

personal and election-related information, which accounts for more than 60% of the US population [5].

As such, if the cloud service does not consider the security of the insider, it can be seen that the amount of data that can be leaked due to one security incident and the magnitude of the damage are not limited.

3 Internal Security Characteristics

Reliability Reliability is the most important in the certification scheme. It is a belief that other entities can assume the same behavior as they expect. This trust can only be applied to some specific functions. The main role of this trust in the authentication framework is to describe the relationship between the entity to be authenticated and the authentication system. The entity to be authenticated must be able to believe that the authentication system produces valid and reliable results.

Multimodality Today, computer and human communication are being done in various ways. In addition to simple keyboards and mice, there are techniques through natural voices, human eyes, and finger movements. This can be one of the ways to authenticate a person's identity as well as the convenience of each person.

Robustness There must be no mistake in authentication. Depending on the environmental factors or the condition of the user on that day, the result should not be different. Unpredictable authentication systems will leave concerns about security threats intact. In the event of an emergency such as a fire, the environment of the server management room should be changed so that the authentication is not possible or the administrator is not mistaken for authentication due to illness.

Reaction speed and ease of use It is also very important whether it is advantageous to use as well as authentication problems. Because many cloud services are used, the server management for problem-solving should be done in a short time. The purpose of HCI technology is to make it easy for users to use the computer. Cloud managers should avoid wasting effort on certification.

By combining the above characteristics, security technology to prevent the threat of malicious insider should be obtained. There is a need to take advantage of the latest HCI technology to improve ease of use and reliability. In addition, there is a need for models that utilize various HCI technologies to improve multimodality.

4   Existing Models

4.1   CERT

CERT is the most commonly used model. It's not specifically used in cloud computing but in all computing that manages servers.

The CERT performs authentication in three steps, as in Fig. 1. The steps can be classified in the order of preventive, detective, and corrective. If a malicious insider is not prevented for the first time, it should at least be detected. Corrective control is the appropriate response to detected insiders. The core of this model is

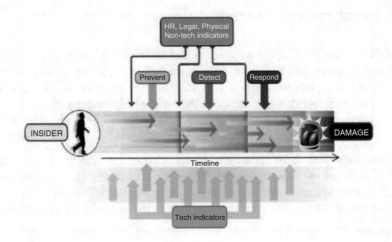

Fig. 1 CERT model diagram [6]

the convergence of technical and nontechnical methods. Policies are integrated into the model to enhance security. In principle, the cert model does not address specific technologies but rather the direction of overall security. Therefore, it is the job of each organization to specify what skills and policies will apply.

4.2 Model-Based Prediction

This model analyzes users' patterns and raises issues when they are not expected behavior. As Fig. 2 shows, the first step is to collect data. This data is obtained through user actions. The problem arises at this stage: noise data. Proper control of the noise data is required. Since this algorithm changes according to the usage environment, a high level of application technology should be used. Once the technology is applied, it will be possible to filter out the wrong insider in a dynamic environment. The model presented in this paper will further strengthen and refine the existing model using HCI technology. Basically, the concept of cert remains the same. In addition, by applying HCI's advanced technology to model-based prediction, data collection for behavior detection will be applied more widely.

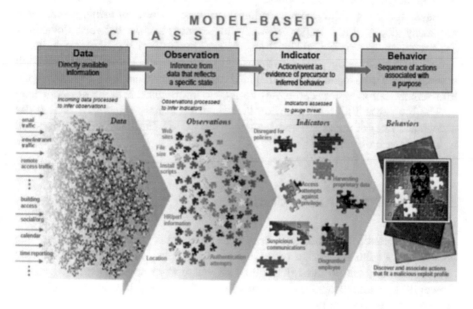

Fig. 2 Model-based prediction [7]

5 HCI Authentication Analysis

The management of the cloud is the role of the server that controls the virtual network. This device authentication should create a multi-authentication-based environment that performs identification and authentication with authorized users. Device authentication is a function basically required to prevent information leakage of customers stored in a cloud server and prevent access to a control system disguised as an authorized user. HCI technology is diverse. What each technology needs is different. Knowing each ability is necessary to make an appropriate model.

5.1 PKI (Public Key Infrastructure)

Symmetric algorithm authentication is a method of mutual authentication between devices by proving an authenticated secret key by mutually exchanging authentication values based on a pre-shared secret key between two devices. This authentication is suitable for small-scale networks with a static structure rather than a large-scale network because of the restriction of sharing a pre-shared secret key for data integrity and confidentiality.

Asymmetric algorithm authentication is suitable for environments where there are many devices in a large network and communication between devices is dynamic. Asymmetric digital signatures are also made up of ID/password authentication, which means that if user information is hacked on a vulnerable website, all authentication information is vulnerable. To solve this problem, an electronic signature authentication system combining biometric information is being developed.

The biometric PKI authentication system using fingerprints has a FAR (false acceptance rate) of about 2% and accuracy of 95% [8]. Further research is needed to increase accuracy.

5.2 ID/Password Authentication

ID/password authentication, well known to the public, is vulnerable to key log attacks and stealth attacks. In order to respond to a key log attack, ID/password can be randomly placed through a virtual keyboard to respond to a key log attack. However, a stealth attack can have a password obtained if a random attack is continuously performed [8].

5.3 Biometrics Authentication

Futurists predict that all identification and authentication of the world is done through biometric information and further that all economic activity can be done in biometrics. In 2015, Alibaba CEO, Jack Ma, announced Alibaba's e-commerce authentication solution and demonstrated e-commerce authentication through face recognition to publicize the idea that the simple payment service using biometric information will be used as a new business model of Alibaba [9].

Biometrics technology has been actively used by the US Department of Homeland Security for terrorist detection since the September 11 attacks and has been using biometrics technology at international airports around the world. The reason why the biometrics technology has become popular all over the world is because the uniqueness of human biometrics information is superior to any other authentication methods and has convenience and simplicity than other authentication methods.

Biotechnology can be divided into physical and behavioral characteristics. Physical characteristics can be divided into face, iris, fingerprint, and vein areas. Behavioral characteristics can be verified by voice, signature, keyboarding habit, and gait. The characteristics of each factor are different as shown in Table 2. A biometric authentication system has strong security, but there is still a risk of counterfeiting. This threat increases even more with a single biometric recognition system.

Currently, fingerprint recognition technology, which is widely used in the field of biometrics recognition, is being used most actively because of its low development cost and excellent security. However, fingerprinting technology also poses a risk of counterfeiting, and research has shown that 2% of the world's population is not fingerprinted [10].

One of the fastest-growing areas of biometric technology is in the fintech (finance technology) sector. Consumers are increasingly interested in easy payment, and as the development speed of smart devices increases, they are used in currency, stock, and finance. In the field of cloud, if it is applied to various authentication methods

Table 2 HCI authentication features

Factor	Type	Action	Factor	Type	Action
PIN code	K	A	Hand geometry	BI	A/P
Password	K	A	Vein recognition	BI	A/P
Voice	BI/BE	A/P	Thermal image	BI/BE	P
Facial	BI	A/P	Behavior patterns	BE	P
Fingerprint	BI	A/P	Electrocardiographic	BI/BE	P

K knowledge, *BI* biometric, *BE* behavior, *A* action, *P* passive that utilize biometric authentication system that enhances security, it can be utilized as an alternative to security measures for device.

Fingerprint recognition technology It has become the most widely used biometrics technology since it is actively used in access control systems, fintech, and mobile. At present, 80% of the biometric market is reported as fingerprint recognition technology [10]. As a result, it has become the most exposed to hacking threats. Especially, it started to be used for self-certification at the time of simple settlement or Internet banking payment, and it has become the most targeted attacker of crackers who are looking for monetary compensation.

Fingerprint forgery can easily be done by the general public, and since the technique is not difficult, there is a problem that the recognition rate and the forgery can be easily made. Especially, it is easy for ordinary people to risk forgery and falsification because it is possible to counterfeit through materials that can be easily obtained. Compared to the counterfeit fingerprint technology that is currently being developed, the corresponding manual and countermeasures are insufficient.

Vascular technology It is a means of authentication using blood vessels. The human body consists of arteries and veins. Arteries vary in number depending on the characteristics of the human body, but the veins are formed during childhood and are not changed afterward. Using this part, we use the second node of the finger which does not change very much, and it irradiates the near-infrared to the target finger. In the vein, hemoglobin has a property of absorbing near infrared rays. Near-infrared rays absorbed from veins leave a black pattern through oxygen and store the characteristics of the pattern, thereby performing authentication of the user. Because of this feature, the possibility of counterfeiting is very low. In addition, it has the advantage of being less influenced by deformation or damage of biometric information [11].

Iris recognition technology The importance of iris recognition is increasing due to the error rate and the risk of fingerprint recognition. The iris recognition technology has the advantage of lowering the error rate than conventional face recognition or fingerprint recognition. The biggest feature of iris recognition is that it is impossible for forgery.

The machine segments the information from the pupillary boundary to the limbus boundary in Fig. 3. Segmentation is the most important step in authentication, where the images the machine gets are different for every person [12]. The recognition speed is less than 1 second, and the error rate (false recognition rate) is low as 10^{-52} because the forgery or damage rate of iris is low [13]. The iris is the most complex and elaborate fibrous tissue in the human body. It is located between the cornea and the crystalline lens and serves to control the amount of light entering the pupil, such as the iris of a camera.

Although the security is high, there is a problem that the error rate is increased due to deformation or various ocular implants (such as circle lens) in the eyes of

Fig. 3 Details of iris anatomy. (Source: Dr. Jan Drewes)

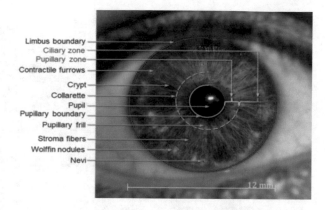

Limbus boundary
Ciliary zone
Pupillary zone
Contractile furrows

Crypt
Collarette
Pupil
Pupillary boundary
Pupillary frill

Stroma fibers
Wolffin nodules

Nevi

12 mm

Fig. 4 Elements for signature authentication. (Source: cs.sabanciuniv.edu)

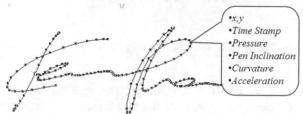

•x,y
•Time Stamp
•Pressure
•Pen Inclination
•Curvature
•Acceleration

the person to be identified. In addition, there are cases of misrecognition due to the illuminance of light [13].

Signature recognition In the signature recognition technology, there is an offline method of recognizing an already created signature and a method of acquiring a signing process online. Of these, a dynamic method is known to be superior in terms of security.

The type of information collected at the time of signing to perform authentication varies from system to system, usually including signature time, speed, number of times the pen has fallen from the paper, and the time at which it occurred [14]. As Fig. 4 shows, divide the signature into parts, and analyze them according to each element.

Facial recognition system Currently, it is widely used for crime investigation and physical access control. In recent years, face recognition has been applied to smartphones and smart TVs, and applications have been expanded to simple payment and financial applications. In recent years, face recognition technology has developed into identification technology for determining who is a person, attribute recognition technology for recognizing face gender and age, and emotion recognition technology for recognizing facial emotion state.

Face recognition technology has been developed for a short time. The probability of misrecognition based on stationary front face photographs lowered from 79% in 1993 to 0.29% in 2010 [15]. As shown in Fig. 5, face recognition technology

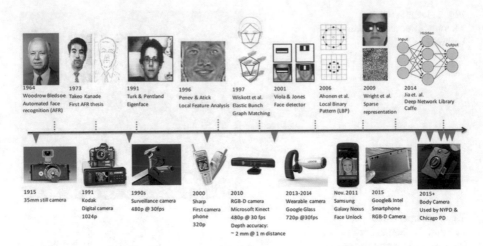

Fig. 5 Major milestones in the history of face recognition. (Source: Researchgate.net)

has been applied to a wide variety of applications such as mobile camera, CCTV, and black box. There are few restrictions on image acquisition compared to other biometric technologies such as fingerprint and iris. Also, people feel less rejection. However, face recognition technology also has a disadvantage in that it has higher risk of forgery and alteration of the subject's face or random change due to plastic surgery than other biometrics technologies, such as fingerprint recognition technology [16].

Voice recognition system The speech recognition system depends on the characteristics of the individual's unique voice. These characteristics are formed by the combination of behavioral and physiological factors. Most speech recognition systems rely on text that a particular word has to be spoken for recognition. Speech recognition technology identifies a person's voice by sampling once every 1/100 second and storing it in the form of a voice frequency graph [17].

Speech recognition systems are the most natural form of biometric technology without physical contact. Convenient, but less reliable. Problems can arise when using speech recognition systems for patients with colds or laryngitis. This technology is the trickiest and most inaccurate.

Hand shape recognition Since the shapes of the hands and fingers are unique to each person, the geometric information obtained by measuring the length and shape of the finger in three dimensions is relatively easy to collect and process. Figure 6 shows what points the machine checks and measures length and width.

However, it is mainly used in places where security is not so important because it is relatively inaccurate. It is used mainly in construction sites or outdoors because it operates stably even in a harsh environment and has less information storage than a system using fingerprint or eye [18].

Fig. 6 Pointing a hand to measure length and width. (Source: uos.ac.kr)

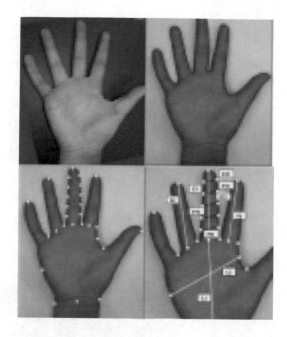

5.4 FIDO (Fast IDentity Online) Authentication

With the development of the IT infrastructure, access to information has also become faster than ever before. The basic authentication methods for the operation and use of the information system service include a password and a PIN number. Passwords are the most common means of authentication, but they can cause security incidents if they are exploited by hacking or intentional attacks, such as malicious code.

Recognition of the problem of passwords has existed for a long time, and many studies have begun to solve them. In 2014, The FIDO alliance has complemented these issues with a technology called FIDO. This technology has separated authentication protocol and method. It is a technology that improves user's convenience while increasing authentication strength without password. Unlike password databases, FIDO stores personally identifying information (PII), such as biometric authentication data, locally on the user's device to protect it. This can overcome the problem of passwords.

As shown in the Fig. 7, FIDO proposes two protocols. First, UAF (universal authentication framework) is a technology that authenticates a user by linking an authentication method provided by a user's device with an online service [19].

The second is the U2F (Universal 2nd Factor) Protocol, which is a protocol that allows strong authentication to be added at user login as the second authentication factor in the online service that uses the existing password. FIDO can distinguish the part of the authentication token that the user authenticates and the authentication

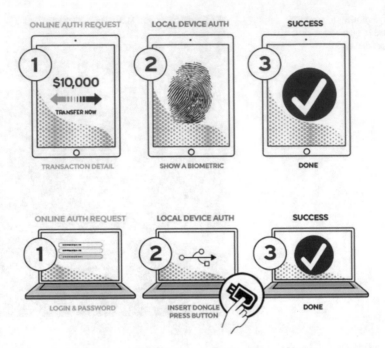

Fig. 7 Multifactor authentication using FIDO. (Source: techxplore.com)

protocol used by the client and the server, thereby diversifying the authentication method without using a password [19].

The user can experience convenience by logging in to the service using bio-based authentication or pattern recognition and can secure more stability than the password by authenticating the user based on the public key between the client and the server. With the introduction of FIDO authentication technology, security of cloud control system can be strengthened, and simple authentication can be realized.

6   Proposed Model

Biometrics is a method for identifying and authenticating authorized users of biometric information based on physical and behavioral characteristics. The previously mentioned iris, vein, fingerprint, and face recognition technologies have advantages and disadvantages depending on the usage environment and the modification of the collected information when using a single authentication, which causes problems in authentication reliability and safety.

Therefore, multiple biometric authentication is a way to overcome the limitation of reliability and safety of single biometric authentication. This helps improve FAR (false acceptance rate) and FRR (false recognition rate).

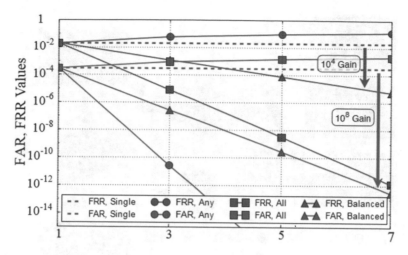

Fig. 8 Gain through the use of multiple recognition systems [20]

As Fig. 8 is seen, Aleksandr Ometov et al. announced that it can reduce the error rate of up to 10^8 when using multiple recognition system [20].

Implementation of multiple authentication systems requires proper sharing of multimodal-based sensor recognition algorithms and efficiency by integration level. Multiple biometric systems require the following requirements [21]:

- Utilizing multiple heterogeneous sensor for the same person
- Utilizing multiple heterogeneous sensor for heterogeneous people
- Utilizing multiple recognition information for the same person
- Utilizing multi-snapshot information of the same person
- Utilizing multiple matching information for the same person

The process of biometrics is shown in Fig. 9. General biometric systems consist of biometric sensors, image processing algorithms, data storage modules, matching algorithms, and decision modules. The biometric data collected by the sensor converts the acquired information into digital data. The system can adopt a plurality of heterogeneous sensors. The imaging processing algorithm extracts significant information from the output image of the sensor and develops a biometric index. The matching algorithm consists of comparing biometric indices based on the generated reference data to generate matching scores and finally using the matching results to determine system level awareness.

By replacing existing knowledge base (password, etc.), authentication with HCI biometric technology, security, and convenience can be improved. Since internal control plays an important role in cloud computing, countermeasures are needed to prevent insider malfunction and system destruction. For this, an authentication model combining two or more biometric technologies is proposed. However, simply combining multiple biometric data can waste unnecessary time. The collected data

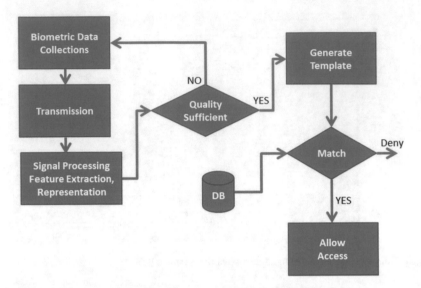

Fig. 9 Biometric authentication process

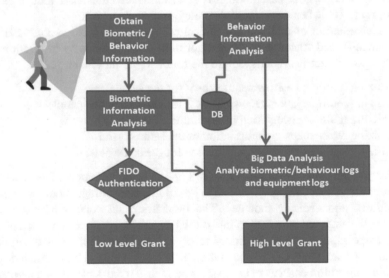

Fig. 10 Multiple biometric authentication process model

are classified according to the characteristics and transmitted to FIDO authentication or biometrics DB.

As shown in Fig. 10, physical authentication is used where the security level is low, taking advantage of the minimization of user inconvenience. However, this is linked to FIDO to further enhance security. Low-level authentication allows access and monitoring of files, but no modification or control is possible.

A higher level of authentication should be handled multiple times with physical authentication plus behavioral or device authentication. This information should be authenticated according to the process in Fig. 10. In addition, the information log is recorded through unstructured data analysis such as classification, clustering, machine learning, and sentiment analysis, which are big data analysis techniques, and authentication is rejected for suspicious behavior.

7 Result and Discussion

Managing an effective cloud network environment requires thorough authentication control. Network attacks in the external environment are becoming increasingly difficult due to the development of technology. However, APT (advanced persistent threat) and insider attacks are relatively vulnerable because they are aimed at human beings. The cloud administrator, who is the target of the authentication, will need not only the user authentication but also the authentication of the device.

It is necessary to introduce and build an integrated intelligent authentication system that detects abnormal signs and analyzes human characteristics through human actions. This paper investigated whether biometric technology of HCI and FIDO could solve the problem and also proposed a model in which the control system is appropriate by combining the above technologies. The standard for cloud administrator privileges is not formalized. Therefore, no model has been derived here that meets the standards and regulations.

There are limitations of biometrics as well as the above problems. Usually biometric information has unchanging uniqueness; it is said to use a lot because it has a strong character. However, in fact, biometric information can be changed by various natural and social factors. Therefore, it is a wrong idea to judge that biometric information is the only information that does not change. One example is that the face type changes with age. Therefore, we need to find ways to overcome these problems. It is necessary to develop an AI system that periodically recognizes that the biometric information update or biometric information difference is a natural change.

8 Conclusion

Reliability and complete security are the most important factors in how the future cloud will evolve. From the first concept of cloud to the present, security concerns have been continuing, and many security incidents have come true. But for this reason, abandoning cloud technology is a waste of its strength and appeal.

Currently, research on cloud security is continuing in various aspects. The results of the research on network and antivirus have achieved not only great quantitative results but also many visible results. However, there is no standardized authentication process model or related policy for insiders. As a result, emerging cloud providers can easily be exposed to security threats with little experience. The currently offered model is CERT. This model will be the most basic concept. To defend against advanced attack techniques, advanced defense techniques are needed.

Although HCI technology was developed for human convenience, its reliability is high enough to be used for human recognition authentication. Among them, the development using biotechnology has been the most remarkable growth in recent years. Using this biometric technology, it would be possible for machines to judge and filter people.

The proposed model is to collect human behavior data through biometric technology and to adjust the authentication stage of a person by probabilistic judgment. Probability algorithms have not been specified, but I believe that future research under these concepts will enable the development of enhanced authentication systems that can be used in practice.

There is no end to security. As the attack method changes, the defense method also needs to be improved. In order for cloud computing, the core technology of the Fourth Industrial Revolution, to be widely utilized, security fears must be addressed first, and research on insider control programs must continue to improve reliability.

References

1. Nayyar, A. (2019). *Handbook of cloud computing: Basic to advance research on the concepts and design of cloud computing* (pp. 10–11). BPB Publications.
2. Baars, T., & Spruit, M. (2012). Designing a secure cloud architecture: The SeCA model. *International Journal of Information Security and Privacy (IJISP), 6*(1), 14–32.
3. Prowell, S., Kraus, R., & Borkin, M. (2010). *Seven deadliest network attacks* (pp. 101–120). Amsterdam: Syngress.
4. Verizon. https://thehackernews.com/2017/07/over-14-million-verizon-customersdata.html
5. DRA. https://thehackernews.com/2017/06/us-voters-data-leaked.html
6. Claycomb, W. R., & Nicoll, A. (2012). Insider threats to cloud computing: Directions for new research challenges. In *IEEE 36th annual Computer Software and Applications Conference (COMPSAC)* (pp. 387–394).
7. Nkosi, L., Tarwireyi, P., & Adigun, M. O. (2013). Insider threat detection model for the cloud. In *Information security for South Africa*.
8. Lozupone, V. (2018). Analyze encryption and public key infrastructure (PKI). *International Journal of Information Management, 38*, 42–44.
9. Jia, K., Kenney, M., Mattila, J., & Seppala, T. (2018). *The application of artificial intelligence at Chinese digital platform giants: Baidu, Alibaba and Tencent*. (ETLA Reports No. 81).
10. Cao, K., & Jain, A. K. (2019). Automated latent fingerprint recognition. *IEEE Transactions on Pattern Analysis and Machine Intelligence, 41*(4), 788.
11. Johnson, M. L. (2004). Biometrics and the threat to civil liberties. *Computer, 37*(4), 90–92.
12. Daugman, J. (2008). Iris recognition. In A. K. Jain, P. Flynn, & A. A. Ross (Eds.), *Handbook of biometrics* (pp. 71–90). Boston: Springer US.
13. Rathgeb, C., & Busch, C. (2017). Improvement of iris recognition based on iris-code bit-error pattern analysis. In *International conference of the Biometrics Special Interest Group (BIOSIG)*.

14. Abikoye, O. C., Mabayoje, M. A., & Ajibade, R. (2011). Offline signature recognition & verification using neural network. *International Journal of Computer Applications, 35*, 44–51.
15. Jain, S., Hu, C., & Aggarwal, J. K. (2011). Facial expression recognition with temporal modeling of shapes. In *ICCV workshops, 2011* (pp. 1642–1649). IEEE.
16. Guo, Y., Zhao, G., & Pietikainen, M. (2012). Dynamic facial expression recognition using longitudinal facial expression atlases. In *ECCV* (pp. 631–644). Springer.
17. van lancjer, D., & Kreiman, J. (1987). Voice discrimination and recognition are separate abilities. *Neuropsychologia, 25*(5), 829–834.
18. Park, G., & Kim, S. (2013). Hand biometric recognition based on fused hand geometry and vascular patterns. *Sensors, 13*, 2895–2910.
19. Fido Alliance. https://fidoalliance.org
20. Ometov, A., Petrov, V., Bezzateev, S., Andreev, S., Koucheryavy, Y., & Gerla, M. (2019). Challenges of multi-factor authentication for securing advanced IoT applications. *IEEE Network, 33*, 82–88.
21. Jain, A. K., Rossand, A., & Prabhakar, S. (2004). An introduction to biometric recognition, IEEE transactions on circuits and systems for video technology. *Special Issue on Image- and Video-Based Biometrics, 14*(1), 4–20.

A Study and Analysis of Issues and Attacks Related to Recommender System

Taushif Anwar and V. Uma

1 Introduction

Recommender systems (RSs) are responsible for providing users with personalized recommendations for services or products. Nowadays, rapid growth of information system and increase in the number of users and data have become big problems in accessing essential information. RS is a widely accepted approach for information overhead and retrieval problem. Through the assistance of RSs, users can effortlessly get interested items from the enormous amount of products. In RSs, user preference data is stored as rating profiles in a database, and they are used in finding their preferences [1].

RSs are widely applied in numerous application domains, including tourism, e-learning, e-commerce, healthcare, sports, etc. RSs are mainly established and designed on the basis of data mining techniques [2], heuristics-based approaches [3], association rule, and similarity measure-based pattern mining techniques [4].

A big challenge for RSs is that data exists in various patterns, viz., user likes, purchased items, most viewed, URL logs, and multimedia elements of web pages. Filtering approaches remove redundant and undesirable information from the large information space and provide only those information that the user is interested. RSs are generally categorized into three types, viz., collaborative filtering (CF), content-based filtering (CBF), and hybrid filtering (HF) as shown in Fig. 1, on the basis of how recommendations are made.

CF approach works depend on both the user's previous ratings and the ratings given by users having similar preferences for an item [5, 6]. CF approaches are classified into two different classes: user-based CF and item-based CF. Mainly, these

T. Anwar (✉) · V. Uma
Department of Computer Science, School of Engineering and Technology,
Pondicherry University, Puducherry, India

© Springer Nature Switzerland AG 2020
S. Paiva, S. Paul (eds.), *Convergence of ICT and Smart Devices for Emerging Applications*, EAI/Springer Innovations in Communication and Computing,
https://doi.org/10.1007/978-3-030-41368-2_7

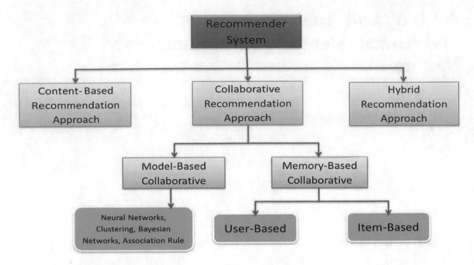

Fig. 1 Types of recommender system

filtering approaches are being used in most of the e-commerce sites and as well as social networking sites. The main disadvantage of this type of recommendation is the cold start problem. Initially, when systems don't have sufficient information for recommending items, cold start problem arises. For effective filtering using CF, a huge amount of information regarding the items is required [7, 8].

CBF recommendation approach works based on user's past preferences and also based on most viewed, liked, bought, and positively ranked items. Basically, this approach initiates from the information filtering and information retrieval domains. CBF system can recommend items as soon as information about items is available. CBF is also used as a supporting system for CF approaches in overcoming some CF issues. In this approach, cold start problem is solved. In CBF, the problem arises when there are potentially few users which may result in very lesser preferences and various unrated items causing sparsity problem. In this situation, finding users with similar interest is a very challenging problem [9].

HF approach endeavors to merge several recommendation approaches for eliminating the cold start, data sparsity, and overspecialization problem [10]. In various ways, different filtering approaches can be merged. HF approach is presented to overcome the traditional RS problems. Netflix uses HF approach. Mostly, HF is used for hybridization techniques such as feature combining, features augmentation, weighted, switching, mixed, and cascade approaches.

RSs are helpful for both users and service providers. RSs are commonly accepted, and they play a significant role in analyzing and observing user's behaviors and generate recommendations as per their interest. This approach also decreases the time and transaction expenses of selecting and searching items in an online shopping domain. They also enhance revenues by means of selling more products by providing users with their most likely subsequent page.

The improvement of information and communication technology (ICT) has transformed the world and has moved us into the Information Age. However, the access and dealing with this huge amount of information are causing precious time losses. Especially, teachers in higher education use the Internet as a medium to seek advice from materials and content for the improvement of the knowledge over the topics. The Internet has very vast services, and sometimes it is miles hard for users to discover the contents easily and rapidly. ICT includes several applications and services such as e-learning, e-health, e-tourism, e-banking, e-government, etc. [11], which play a remarkable role in the development of new media ways for interaction among people. Mobile learning is one of the trends that expand exponentially, even in developing countries. In general, if students have a multimedia mobile and Internet connection, even if the school lacks ICT support, students can easily have access to online materials. Handling large amount of information and getting accurate information within the stipulated time are a difficult problem. In this scenario, the recommender system helps us and provides accurate information within a short span of time using our past and demographic information.

The remainder of the paper is organized as follows. Section 2 describes the recommender system model, followed by the explanations regarding collaborative and content-based filtering. Section 3 describes the matrices for recommendations. Section 4 presents issues related to recommender systems. Section 5 reviews some related work that was carried on issues of the recommender system. Section 6 presents a comparative analysis of collaborative and content-based filtering. Section 7 describes common attacks related to recommender system. Section 8 provides details about applications of ICT in context of recommendation. Section 9 evaluates the proposed approach using MovieLens datasets, and Sect. 10 concludes the book chapter.

2 Recommender System Model

Recommender system suggests items to users based on their interests and behaviors [12]. Figure 2 shows the simple model of a traditional recommender process. This figure shows that user's profiles and item features are obtained by profile and item manipulation. Profile–item matching is performed and top-N-items are suggested based on filtering. CF works based on the profile and ratings given by users. CBF works based on user past preferences and profile attributes of users.

3 Metrics for Recommendation

Ranking of items plays important role in recommender systems. Ranking measures are used widely when a list of ordered items given to the user is on the basis of

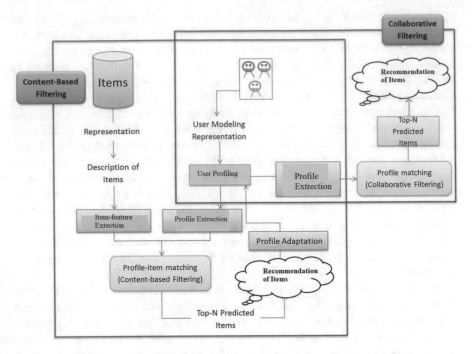

Fig. 2 Simple model of recommender process

Table 1 Confusion matrix

	Recommended/ positive prediction	Not recommended/ negative prediction	Total
Used (relevant)	True positive (TP)	False negative (FN)	Total relevant
Unused (irrelevant)	False positive (FP)	True negative (TN)	Total irrelevant

their interest. The order format can be as follows: the least relevant items are at the bottom, and the most suitable and similar items are at the top.

In the field of information filtering, selecting good items from the background of bad items can be viewed as refining relevant (used) and irrelevant (unused) items. So, recommender system provides the items that users may like and the standard way to measure. It is based on classification metrics like precision, accuracy, recall, and false positive rate [13]. These metrics have been widely used in several domains to incorporate recommendations as indicated in Table 1.

The classification metrics like accuracy, precision, recall, and specificity are briefly explained as below.

3.1 Accuracy

Accuracy as in Eq. 1 indicates how close a calculated value is to the real (true) value.

$$Accuracy = \frac{(TP + TN)}{(TP + TN + FP + FN)} \tag{1}$$

3.2 Precision

Precision is defined as how close the measured values are to each other, or it may denote the relevancy of the selected items [14]. For example, when an e-commerce site recommends 20 mobiles and only 10 mobiles are relevant while failing to return 30 additional relevant mobiles, its precision as per Eq. 2 is 0.5 (TP = 10, FP = 10).

$$Precision = \frac{TP}{(TP + FP)} \tag{2}$$

3.3 Recall

Recall denotes the number of relevant or positive items the model returns. It is also known as the true positive rate or sensitivity [15]. For example, when e-commerce sites recommend 20 mobiles and if only 10 were relevant while failing to return 30 additional relevant mobiles, then its recall as per Eq. 3 is 0.25 (TP = 10, FN = 30).

$$Recall = \frac{TP}{(TP + FN)} \tag{3}$$

3.4 Specificity

Specificity is also called true negative rate. It is used for measuring the number of negative items that are correctly identified. The said formula is in Eq. 4.

$$Specificity = \frac{TN}{TN + FP} \tag{4}$$

4   Issues in Recommender System

4.1   Types of Issues in Collaborative Filtering

Data Sparsity Problems

The problem arises when there are potentially few users which may result in very lesser preferences and various unrated items causing sparsity problem. In this situation, finding users with similar interest is a very challenging problem [15]. The system becomes very ineffective, and it reduces the neighbor transitivity and coverage. Neighbor transitivity problem occurs through sparse databases in which users with related interest identification are difficult if they have not given the rating to the same items.

Cold Start Problem

It is also known as ramp-up or start-up problem. There are mainly two related issues in cold start problem: new user and new item problem. Cold start problem arises when the new user and new item enter the system and when enough data/information has not yet been gathered [16]. This is because these systems require a huge volume of data regarding users or items for making appropriate recommendations. New items cannot be suggested by the system until some users rate them. Cold start problems can be solved by (a) explicitly asking the users about his interest, (b) recommending item based on demographics, and (c) asking the user to rate some items at the beginning information.

Scalability

If the number of users, items, and evaluations is too large for the system, the time taken for processing real-time relations may be very high leading to very high response time and may require some resources that are unavailable. This problem exists in both collaborative and content-based filtering approaches. For example, Amazon.com has approximately 20 million users and recommends more than 18 million items. For overcoming this problem and speeding up recommendations, dimensionality reduction, clustering, matrix factorization, and using Bayesian network were used. Generally single-value decomposition (SVD) [17] is used for dimensionality reduction. Clustering increases the performance of recommendation but decreases the accuracy because clustering algorithm finds users/items in small clusters instead of the entire database. Matrix factorization methods are not suitable for e-commerce recommendations with big datasets.

Gray Sheep

Gray sheep problem occurs if users have the rare taste. In this case, the recommendation may not be accurate, as there are no close neighbors. This problem can be resolved by pure content-based filtering in which items are recommended by manipulation of the description of items and user personal profile [18]. For example, if the system identifies that users are interested in oceanography and technology, the related items will be easily recommended to the user, even if unpopular items are considered.

Early Rater Problem

Collaborative filtering system cannot provide the recommendation about newly added items because there is no rating available at that time. When the user starts to give the rating and even if the item has gathered enough rating, providing accurate recommendation is difficult. Similarly, recommendations will not be accurate for new users who have rated only a few items [19].

Synonymy

This problem occurs when similar or the same items have various names and the RS fails to predict accurate items [20]. In these situations, the RS cannot find whether the terms presented denote same items or different items. For example, memory-based BF system will treat "Hollywood film" and "Hollywood movie" differently. To resolve the synonymy problem, different approaches including SVD (single-value decomposition, ontologies, and LSI (latent semantic indexing) are used.

Shilling Attack

It mainly concerns with collaborative filtering but has a minor threat compared to the item-based CF approach. This problem arises mainly when a malicious user enters the system and provides negative rating for their competitors and a lot of positive rating for their own items. Its main target is to either decrease or increase item's popularity [21]. This type of attack decreases the quality and performance of the recommender system and can become a barrier to the trust of the recommender [22].

Latency Problem

Latency problem is mainly related to cold start new item problem. This problem arises when frequently new items are added to the database. Collaborative filtering approach suffers from this problem because recommender systems recommend only those items which have been sufficiently rated and hence the newly added items will not be recommended till sufficient rating has been gathered. The waiting time can be reduced by using content-based filtering, but it may lead to overspecialization. To overcome this problem, the category-based approach merged with the user stereotype can be used [23].

Privacy

For providing better recommendation, i.e., what user wants or needs, recommender system generally relies on the large amount of user data gathered from user interest [24]. It may lead to data security and privacy concern issues. User data either collected explicitly (product rating, comments, likes) or implicitly (purchase history) also brings about privacy concern as this data also contains some demographic data which can point to singular identity (social security number, email), product-related information, and footprints that the user leaves online regarding the web browsing information such as purchasing and search habits.

User's data is stored generally in the centralized repository in collaborative filtering which may result in data misuse. For overcoming this, cryptographic techniques and semantic web technologies (ontologies) with natural language processing (NLP) are used to diminish nonessential exposure of information [25].

4.2   Types of Issues in Content-Based Filtering

Limited Content Analysis

Limited content analysis problem arises when two different types of products are defined using the same set of attributes. Hence, they become very difficult to identify. The availability of content is limited, and hence it leads to the overspecialization problem [26]. Items are generally represented and selected based on subjective attributes. To get a satisfactory set of attributes, the attributes should be assigned manually, or content is to be parsed automatically. Automatic parsing is very easy for text features but not for videos and images.

Overspecialization

Content-based filtering approach will not select or recommend items if the previous user activities do not provide any evidence. Sometimes, the user attempts to try something related to new and novel items, but the system would never provide the details. It prevents the user from knowing other options and discovering new items, and so diversity of recommender system decreases. Diversity is a desirable feature of recommendation approach.

For recommending serendipitous items (accidentally suggests items unexpected but useful) [27] and novel items along with known items, some randomness can be added.

Learning Algorithm Problem

Choosing a learning algorithm is also an issue because the efficiency of a learning method does play a vital role in decision-making. Storage space and computational complexity of the algorithm can also become an issue in managing user profiles. Genetic algorithm and neural network are slower compared to other learning methods because several iterations are required to determine the relevancy of the document. Relevance feedback and Bayesian classifier are used for increasing the speed and performance.

5 Related Work Regarding Recommender System Issues

The root of the RSs began with research papers on CF by W. Hill et al. [28] and Resnick et al. [29]. Numerous methods such as clustering, regression, decision tree, neural network, and k-NN are applied in RSs. There are some issues such as shilling attack, data sparsity, cold start, scalability, gray sheep, latency, and privacy problems which are faced by RSs.

Sh. Asadi et al. [30] presented a course recommender system using clustering and fuzzy association rule mining. In general, through ICT, student is getting the course list, and selecting an appropriate course is a difficult problem. The main goal of this work is to develop a course recommender system that takes the student's information into account to suggest proper courses. Clustering is used for finding the students having similar skills, interest, and behavior. Fuzzy association rule mining is applied for analyzing patterns in course selections by students as well as the associations between them. The mined rule helps in decision-making regarding course selection.

Some researchers like Yongfeng Qian et al. [31] presented an emotion-aware RS using hybrid information fusion. The author tried to consider user's emotional changes, which play an imperative task in consumption activity. For hybrid information fusion, three types of information are analyzed, viz., user social network data, user rating data, and user reviews. The experimental evaluation is done by watercress datasets, and results prove that the presented model significantly increases the recommendation accuracy and provides a higher prediction rating.

Jingwei Xu et al. [32] presented a solution for the cold start problem by using generic RAPARE strategy. This allowed the system to provide faster recommendation with linear scalability and also gives special treatment for both new users and new items problems. MovieLens and real dataset are used for experimental testing.

Andy Yuan Xue et al. [33] have given a solution for data sparsity problem. The solution is related to destination prediction using SubSyn (Sub-Trajectory Synthesis), SubSynE (E for Efficiency), and SubSynEA (A for Accuracy), and the procedure is expressed on the basis of the Markov model. For improvement in accuracy, this algorithm uses the space partitioning techniques and second-order Markov model. In this paper, real-world T-drive taxi trajectory project dataset is used for experimental testing. The result shows improvement in prediction accuracy and run time accuracy.

Ke Ji et al. [34] presented a solution for scalability problem wherein the recommendation is made on the basis of tag–keyword relation matrix as a replacement for user–item matrix. Three-factor matrix factorization and neighborhood method are used for building the tag–keyword relation matrix on the basis of observed rating. For experimental testing, KDD Cup 2012 real dataset is used. Through this, increase in scalability is achieved on the incremental processing of new data.

Benjamin Gras et al. [35] proposed a solution for gray sheep users (GSU) problem wherein new distribution-based GSU identification technique is used. The MovieLens dataset is used for experimental testing. The result shows that accuracy achieved is slightly higher.

Karthikeyan P. et al. [36] presented solutions for shilling attack problem where discrete wavelet transform is used to perform the rating series and support vector machine is used for classification. The MovieLens dataset is used for experimental testing. Specificity (true negative rate) and sensitivity (recall) around 90% and precision over 90% are achieved using the proposed technique.

Shahriar Badsha et al. [37] proposed solutions for privacy problem wherein the main motive is to hide user's private information from the RSs. In this work, privacy-preserving protocol and cryptography-based approach are used. GroupLens dataset is used for experimental testing. In this paper, experimental result shows that other users and servers cannot learn about users and accuracy is also maintained (Table 2).

6 Comparative Analysis of Collaborative and Content-Based Filtering Approaches

From Table 3, it is found that the data sparsity problem is present in both collaborative filtering and content-based filtering. New user problem is present in both the filtering approaches, but new item problem has been solved in CBF. Scalability issue is present in both, but gray sheep problem is solved in CBF. Early rater problem is present in CF but solved in CBF. Synonymy issue is present in both, and shilling attack is present in CF and solved in CBF. Latency problem is present in CF but partially solved in CBF. The limited content analysis is present in CBF but partially solved in CF. Overspecialization and learning algorithm problems are present in both filtering approaches.

Table 2 Related work table

	Authors	Issues targeted	Filtering methods	Techniques	Datasets used	Advantages
1	Jingwei Xu et al. [32]	Cold start problem	Content-based filtering	Generic RAPARE strategy applied on both matrix factorization-based (RAPARE-MF) and neighborhood-based (RAPARE-KNN) collaborative filtering	MovieLens and EachMovie	RAPARE-MF (instantiating with matrix factorization method) can provide fast recommendations with linear scalability
2	Taushif Anwar et al. [7]	Cold start problem	Collaborative filtering	Sequential pattern mining and rule mining are applied	MovieLens and Book dataset from github	Recommendation accuracy improved and sparsity problem alleviated
3	Fernández et al. [38]	Data sparsity problem	Collaborative filtering	Cross domain recommendation and knowledge representation technique	Facebook likes data for movie music and book	Recommendation accuracy improved as well as diversified recommendation provided
4	Ke Ji et al. [34]	Scalability problem with tags and keywords	Collaborative filtering	Tag-keyword relation matrix is used instead of the user–item rating matrix; novel neighborhood approach is used for building the tag-keyword relation matrix	Real dataset published by KDD Cup 2012	Increases the scalability on incremental processing of new data
5	Benjamin Gras et al. [35]	Gray sheep users (GSU) problem	Collaborative filtering	New distribution-based gray sheep user's identification technique used that belongs to the distribution-based class of outlier detection	MovieLens dataset	Accuracy is slightly higher, especially when a larger number of GSU is selected
6	Karthikeyan. P et al. [36]	Shilling attack	Collaborative filtering	Discrete wavelet transform (performed on the rating series) and support vector machine (for classification) are used	MovieLens 100 K dataset	Specificity (true negative rate) around 90%, sensitivity (recall) is maintained at 90%, and precision over 90% is achieved
7	Mikael Sollenborn et al. [39]	Latency problem	Collaborative filtering	Category-based filtering is used and integrated with clustering and user modeling	News and adverts dataset (50 news and 70 adverts, 5 different hypothetical users with their interest)	The latency problem is reduced in highly dynamic domains, and clustered user data enables quicker response times
8	Qiang Tang et al. [40]	Privacy	Collaborative filtering	A new decentralized single prediction protocol used	Twitter datasets	Achieves better accuracy than some baseline algorithm. Security of protocol is guaranteed by the underlying homomorphic encryption scheme

Table 3 Comparative analysis of content-based and collaborative filtering approaches

	Issue name	Collaborative filtering	Content-based filtering
1	Data sparsity	✓	✓
2	New item problem (cold start problem)	✓	✗
3	New user problem (cold start problem)	✓	✓
4	Scalability	✓	✓
5	Gray sheep	✓	✗
6	Early rater problem	✓	✗
7	Synonymy	✓	✓
8	Shilling sttack	✓	✗
9	Latency problem	✓	Partially solved
10	Privacy	✓	✓
11	Limited content analysis	✓	✓
12	Overspecialization	✓	✓
13	Learning algorithm problem	✓	✓

7 Common Attacks Related to Recommender System

In RS, recommending some bogus item by an attacker is called attack. Attacks are classified by how attackers create profile. Generally, shilling attack is classified as push attack and nuke attack. In push attack, attackers provide the highest rating to targeted items for increasing the popularity, and in the nuke attack, attackers give the lowest rating to target items for decreasing the popularity. We are mainly focusing on push attacks such as random attack, average attack, probe attack, bandwagon attack, reverse bandwagon attack, segment attack, and love–hate attack.

7.1 Random Attack

The random attack is caused due to low information and low knowledge [36]. Attack-generated profile is chosen randomly, and it is rated based on the mean rating and standard deviation of the system. The targeted sets of items are rated with maximum and minimum values depending on the nature of the attack, i.e., push or nuke, for example, if the rating is between 1 and 5, where 1 and 5 correspond to disliked and liked items, respectively. Implementation of this attack is generally easy, but effectiveness is poor.

7.2 Average Attack

This attack is complicated than other attacks because it requires more information about the system, recommendation algorithm, and datasets for practical implementation [41]. Generally, it uses particular average rating of the individual item rather than the global average of the system. For attacking, an attacker randomly chooses the items and uses the standard deviation of the particular item and normal distribution with mean.

7.3 Bandwagon

Bandwagon attack is also called as popular attack. Items which are rated by various users are known as popular items. Here, the attacker gets benefits out of Zipf's low distribution of popularity and thereby results in biased outlines that include the most famous items. Hence, there is a high probability that attackers switch to the role of real users [36].

7.4 Reverse Bandwagon Attack

It is an extension of the bandwagon attack. Generally, in this type of attack, the chosen items are to be less rated by various users. The attack gives less rating to these items and the goal item. This technique is related to widely disliked items with the goal item, raising the chance that the system will produce the less-predicted rating for them [42].

7.5 Segment Attack

In segment attack, the attacker changes the goal of attacking from the whole user set to only a section of portraits with a particular interest. The attacks on the segmented way are usually considered more robust and also have a vital impact on the item-based algorithm. It is also more effective on the user-based algorithm [36].

7.6 Probe Attack

In this attack, attackers give the rating to a small number of selected items and can gradually develop an attack profile which will nearly match ratings given by the real users of the system. So, there is a chance of more similarities between real and attack profiles. Less domain knowledge is needed as compared to other attacks.

7.7   Love–Hate Attack

In this attack, knowledge requirements are absent. It works based on the attack profile where the goal item gets less rating while filler items get more rating. Especially, this profile generation approach is much powerful in a nuke attack [43].

The main goal of recognizing the attack profile and eliminate them from the RS before producing the recommendation. There are mainly two types of approaches, namely, item-based and profile-based. Item-based approach selects items whose predictions are beyond the boundaries. This approach depends upon the statistical process control (SPC) technique, where anomaly ratings detect for each items. The boundary created the lower and upper limits which are estimated by two horizontal lines with the use of historical ratings. This approach unfortunately alerts of the presence of the attack, but it is unable to spot the culpable rater. The profile-based approach can be classified into two classes, namely, principal component analysis-based method and classification-based approach. Principal component analysis-based method transforms the user–item matrix to a hyperplane, then it will be identified as an attacker. The classification-based approach exploits a model that was built previously to predict whether a new profile is a genuine user or an attacker.

8   Application of ICT in Context of Recommendation

ICT includes several applications and services, namely, videoconferencing and distance learning. ICT plays a remarkable role in the development of modern ways of interaction among people.

ICTs are playing an important role in the education field. Teachers use ICTs to participate in the online conference and search for teaching materials that can aid in teaching. For students, ICT is used as a reference tool. In the e-learning system, students can access class notes, submit assignments, and also join a discussion group with the help of ICTs [44]. Nowadays, we have a large number of books, articles, materials, etc. in the Internet. So, finding the right materials at the right time is difficult. RS helps us in finding the right materials quickly [45]. Through RS, we can save our time, effort, and money.

In the banking sector, ICTs play a remarkable role and help in controlling the entire banking system which involves electronic banking services. Users use ICTs to make transactions 24*7 (anytime anywhere) and save their time as well as money. For controlling the whole banking system, bank administrators use ICTs. In the field of banking, various types of schemes exist. Selecting suitable schemes is a difficult problem. In this case, RS recommends schemes according to user needs, and it helps to get more benefits [46].

Selling and buying goods using the Internet is known as e-commerce. E-commerce makes buying and selling activities easier and faster. In an e-commerce sector, suppliers, employees, and customers get benefits from the usage of ICT. Buyers use ICTs to connect online with suppliers to purchase products, and suppliers use ICTs to keep track of their transactions [47]. Selecting an accurate product from a large number of products is a difficult task. In this case, the recommender system plays an important role. RS recommends products based on past buying behavior or demographic information of the user.

ICT applications had been precious resources inside the medical area. They assist the green exchange of facts among health professionals. They facilitate the transfer of patient records and can improve the quality of care provided by health professionals [11]. Ultrasound, CT scan, and ECG are achieved with the aid of ICTs in prognosis of different diseases. The advantage of combining information and communication technologies (ICTs) with recommender systems and health sector is apparent as not only patients but also healthcare personnel and the fitness care system are benefited. In this field, RS helps the doctor and recommends what types of tests are required and what types of diagnoses are required based on patient past information. Through this we can save money, time, and lives also.

9 Proposed System Architecture

In this proposed recommender system, recommendation is done using collaborative filtering and intention mining. For finding similarity, cosine similarity is used. After that prediction matrix is generated. The recommendation is done using intention merged with the prediction matrix [45]. The proposed approach will overcome new user problem and improve recommendation accuracy. The inferred intention will also help the RS to enhance the user experiences and provide better services.

Proposed recommender system is split into five modules:

- Data preprocessing
    In data preprocessing, data is transformed into efficient and useful format, and also missing values are identified.
- Calculating similarity between items and users using cosine similarity techniques
    Similarity is calculated using cosine similarity in which more similar items have more chance of recommendation.
- Finding prediction matrix
    The rating of items is calculated using prediction matrix. Through the prediction matrix, missing rating is also predicted.
- Identifying the intentions of the user
    From the user profile, intentions such as age, gender, and time are identified.
- Recommendation based on collaborative filtering and intentions
    Combining collaborative filtering and intentions will increase the recommendation accuracy.

IM is a technique that aims to identify user's intentions from their contextual information or historical interactions. It has been widely applied and studied in various research areas such as web search, social media, multimedia, robotics, etc. The intentions of the users are correlated with user–item interactions, user's

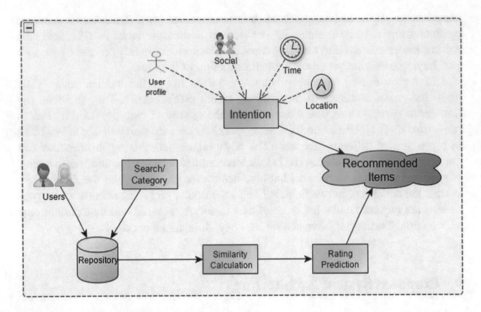

Fig. 3 Proposed model for recommendation using collaborative filtering and intention mining

social influence, and user's social influence. This proposed approach will result in improved recommendation accuracy because it gathers intentions of the user from various sources of raw data [48].

9.1 Datasets and Result Discussion

In this work, for generating intentions, we have used the publicly obtainable dataset for movie domain from MovieLens 1M datasets which contain 6040 users, 3900 movies, and 1,000,209 ratings. From the MovieLens datasets, we have taken users dataset (user ID, gender, age), rating datasets (user ID, movie ID, rating), and movies dataset (movie ID, title, genres).

In this proposed approach, we tried to generate user intention based on the category of movies using the gender and age of the user. At first, we identified the number of male and female users based on age. Age group is divided into seven categories (under 18, 18–24, 25–34, 35–44, 45–49, 50–55, 56+). Figure 4 shows that people in age group 25–34 gave more ratings in both genders. Figure 5 shows the percentage of male users in different age groups based on movie genres. Figure 6 shows the percentage of female users in different age group based on movie genres. The ratings and the intentions present in the datasets will increase the recommendation accuracy.

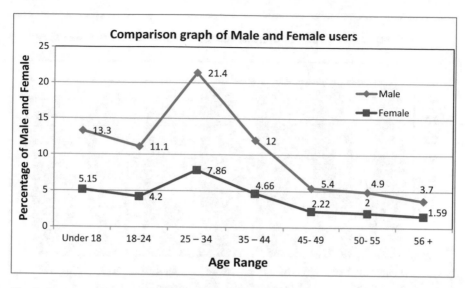

Fig. 4 Comparison graph of male and female based on age

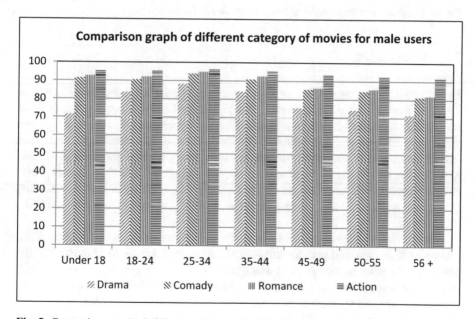

Fig. 5 Comparison graph of different category of movies for male users

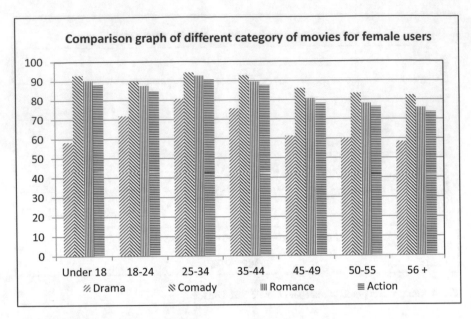

Fig. 6 Comparison graph of different category of movies for female users

10 Conclusion

In the field of recommender system, much research has been done over the last few decades. This book chapter has introduced the RS approaches such as collaborative, content-based, and hybrid filtering approaches that are mainly used for recommendations. Each approach has its own issues and challenges, and much research is being done actively to solve them. We have also surveyed the issues and challenges related to collaborative and content-based filtering. In this work, literature analysis is provided on the basis of issues, techniques used for solving these issues, datasets used for experimental testing, and advantages gained by the researcher.

Various interesting approaches have been developed in this research field. However, there are various challenges and issues that are to be handled. Firstly, there is a demand to analyze and distinguish the various approaches of RS with a standard evaluation platform and to develop and embellish plan guidelines. Secondly, purpose of the research such as reducing and alleviation of the data sparsity, the new item or the new user (cold start), scalability, gray sheep, early rater, synonymy, shilling attack, and privacy problems of recommendations are still under-explored. The chapter concludes by highlighting the issues, challenges, and application of recommender systems.

References

1. Quadrana, M., Cremonesi, P., & Jannach, D. (2018). Sequence-aware recommender systems. *ACM Computing Surveys (CSUR), 51*(4), 66.
2. Witten, I. H., Frank, E., Hall, M. A., & Pal, C. J. (2016). *Data mining: Practical machine learning tools and techniques*. Cambridge, MA: Morgan Kaufmann.
3. Yu, Z., Xu, H., Yang, Z., & Guo, B. (2016). Personalized travel package with multi-point-of-interest recommendation based on crowdsourced user footprints. *IEEE Transactions on Human-Machine Systems, 46*(1), 151–158.
4. Tarus, J. K., Niu, Z., & Kalui, D. (2018). A hybrid recommender system for e-learning based on context awareness and sequential pattern mining. *Soft Computing, 22*(8), 2449–2461.
5. Kumar, P., & Thakur, R. S. (2018). Recommendation system techniques and related issues: A survey. *International Journal of Information Technology, 10*(4), 495–501.
6. Anwar, T., & Uma, V. (2019). MRec-CRM: Movie recommendation based on collaborative filtering and rule mining approach, In *2019 international conference on Smart Structures and Systems (ICSSS)*, Chennai, India, 2019, pp. 1–5. https://doi.org/10.1109/ICSSS.2019.8882864
7. Anwar, T., & Uma, V. (2019). CD-SPM: Cross-domain book recommendation using sequential pattern mining and rule mining. *Journal of King Saud University-Computer and Information Sciences*. https://doi.org/10.1016/j.jksuci.2019.01.012
8. Kumar, P., Kumar, V., & Thakur, R. S. (2019). A new approach for rating prediction system using collaborative filtering. *Iran Journal of Computer Science, 2*(2), 81–87.
9. Kant, S., & Mahara, T. (2018). Merging user and item based collaborative filtering to alleviate data sparsity. *International Journal of System Assurance Engineering and Management, 9*(1), 173–179.
10. Anwar, T., & Uma, V. (2019). A review of recommender system and related dimensions. In *Data, engineering and applications* (pp. 3–10). Singapore: Springer.
11. Kohno, R., Kobayashi, T., Sugimoto, C., Kinjo, Y., Hämäläinen, M., & Iinatti, J. (2019). Medical healthcare network platform and big data analysis based on integrated ICT and data science with regulatory science. *IEICE Transactions on Communications, 102*(6), 1078–1087.
12. Ricci, F., Rokach, L., & Shapira, B. (2015). Recommender systems: Introduction and challenges. In *Recommender systems handbook* (pp. 1–34). Boston: Springer.
13. Li, C., et al. (2017). Deep speaker: an end-to-end neural speaker embedding system. *arXiv preprint arXiv:1705.02304*.
14. Tarus, J. K., Niu, Z., & Yousif, A. (2017). A hybrid knowledge-based recommender system for e-learning based on ontology and sequential pattern mining. *Future Generation Computer Systems, 72*, 37–48.
15. Al-Bashiri, H., Abdulgabber, M. A., Awanis, R., & Norazuwa, S. (2018). A developed collaborative filtering similarity method to improve the accuracy of recommendations under data sparsity. *International Journal of Advanced Computer Science and Applications (IJACSA), 9*(4), 135–142.
16. Zhu, Y., et al. (2019). Addressing the item cold-start problem by attribute-driven active learning. *IEEE Transactions on Knowledge and Data Engineering*, 1.
17. Chai, Z., Li, Y., Han, Y., & Zhu, S. (2019). Recommendation system based on singular value decomposition and multi-objective immune optimization. *IEEE Access, 7*, 6060–6071.
18. Ghazanfar, M. A., & Prügel-Bennett, A. (2014). Leveraging clustering approaches to solve the gray-sheep users problem in recommender systems. *Expert Systems with Applications, 41*(7), 3261–3275.
19. Rashid, A. M., Karypis, G., & Riedl, J. (2008). Learning preferences of new users in recommender systems: An information theoretic approach. *Acm Sigkdd Explorations Newsletter, 10*(2), 90–100.
20. Isinkaye, F., Folajimi, Y., & Ojokoh, B. (2015). Recommendation systems: Principles, methods and evaluation. *Egyptian Informatics Journal, 16*(3), 261–273.

21. Chen, K., Chan, P. P., Zhang, F., & Li, Q. (2019). Shilling attack based on item popularity and rated item correlation against collaborative filtering. *International Journal of Machine Learning and Cybernetics, 10*(7), 1833–1845.
22. Samaiya, N., Raghuwanshi, S. K., & Pateriya, R. (2019). Shilling attack detection in recommender system using PCA and SVM. In *Emerging technologies in data mining and information security* (pp. 629–637). Singapore: Springer.
23. Khusro, S., Ali, Z., & Ullah, I. (2016). Recommender systems: Issues, challenges, and research opportunities. In *Information science and applications (ICISA) 2016* (pp. 1179–1189). Singapore: Springer.
24. Cheng, H.-T., et al. (2016). Wide & deep learning for recommender systems. In *Proceedings of the 1st workshop on Deep Learning for Recommender Systems* (pp. 7–10). ACM.
25. Heupel, M., Fischer, L., Bourimi, M., & Scerri, S. (2015). Ontology-enabled access control and privacy recommendations. In *Mining, modeling, and recommending 'things' in social media* (pp. 35–54). Cham: Springer.
26. Flores-Parra, J. M., Castanon-Puga, M., Martinez, L.-G., Rosales-Cisneros, R., & Gaxiola-Pacheco, C. (2017). Towards recommendation system for work-group formation using social network analysis approach. In *Proceedings of the World Congress on Engineering and Computer Science* (pp. 508–513).
27. Ge, M., Delgado-Battenfeld, C., & Jannach, D. (2010). Beyond accuracy: evaluating recommender systems by coverage and serendipity. In *Proceedings of the fourth ACM conference on Recommender Systems* (pp. 257–260). ACM.
28. Hill, W., Stead, L., Rosenstein, M., & Furnas, G. (1995). Recommending and evaluating choices in a virtual community of use. In *Proceedings of the SIGCHI conference on Human Factors in Computing Systems* (pp. 194–201). ACM Press/Addison-Wesley Publishing Co.
29. Resnick, P., Iacovou, N., Suchak, M., Bergstrom, P., & Riedl, J. (1994). GroupLens: An open architecture for collaborative filtering of netnews. In *Proceedings of the 1994 ACM conference on Computer Supported Cooperative Work* (pp. 175–186). ACM.
30. Asadi, S., & Shokrollahi, Z. (2019). Developing a course recommender by combining clustering and fuzzy association rules. *Journal of AI and Data Mining, 7*(2), 249–262.
31. Qian, Y., Zhang, Y., Ma, X., Yu, H., & Peng, L. (2019). EARS: Emotion-aware recommender system based on hybrid information fusion. *Information Fusion, 46*, 141–146.
32. Xu, J., Yao, Y., Tong, H., Tao, X., & Lu, J. (2017). R a P are: A generic strategy for cold-start rating prediction problem. *IEEE Transactions on Knowledge and Data Engineering, 29*(6), 1296–1309.
33. Xue, A. Y., Qi, J., Xie, X., Zhang, R., Huang, J., & Li, Y. (2015). Solving the data sparsity problem in destination prediction. *The VLDB Journal—The International Journal on Very Large Data Bases, 24*(2), 219–243.
34. Ji, K., & Shen, H. (2015). Addressing cold-start: Scalable recommendation with tags and keywords. *Knowledge-Based Systems, 83*, 42–50.
35. Gras, B., Brun, A., & Boyer, A. (2017). Can matrix factorization improve the accuracy of recommendations provided to grey sheep users? In *13th international conference on Web Information Systems and Technologies (WEBIST)* (pp. 88–96).
36. Karthikeyan, P., Selvi, S. T., Neeraja, G., Deepika, R., Vincent, A., & Abinaya, V. (2017). Prevention of shilling attack in recommender systems using discrete wavelet transform and support vector machine. In *2016 eighth international conference on Advanced Computing (ICoAC)* (pp. 99–104). IEEE.
37. Badsha, S., Yi, X., Khalil, I., & Bertino, E. (2017). Privacy preserving user-based recommender system. In *2017 IEEE 37th international conference on Distributed Computing Systems (ICDCS)* (pp. 1074–1083). IEEE.
38. Fernández-Tobías, I., Cantador, I., Tomeo, P., Anelli, V. W., & Di Noia, T. (2019). Addressing the user cold start with cross-domain collaborative filtering: Exploiting item metadata in matrix factorization. *User Modeling and User-Adapted Interaction, 29*(2), 443–486.

39. Sollenborn, M., & Funk, P. (2002). Category-based filtering in recommender systems for improved performance in dynamic domains. In *International conference on Adaptive Hypermedia and Adaptive Web-Based Systems* (pp. 436–439). Springer.
40. Tang, Q., & Wang, J. (2018). Privacy-preserving friendship-based recommender systems. *IEEE Transactions on Dependable and Secure Computing, 15*(5), 784–796.
41. Zhou, W., Wen, J., Xiong, Q., Gao, M., & Zeng, J. (2016). SVM-TIA a shilling attack detection method based on SVM and target item analysis in recommender systems. *Neurocomputing, 210*, 197–205.
42. Zhang, F. (2009). Reverse bandwagon profile inject attack against recommender systems. In *2009 second international symposium on Computational Intelligence and Design* (pp. 15–18). IEEE.
43. Burke, R., Mobasher, B., Williams, C., & Bhaumik, R. (2006). Classification features for attack detection in collaborative recommender systems. In *Proceedings of the 12th ACM SIGKDD international conference on Knowledge Discovery and Data Mining* (pp. 542–547). ACM.
44. Cerna, M. (2019). Modified recommender system model for the utilized eLearning platform. *Journal of Computers in Education*, 1–25.
45. Nakayama, K., Yamada, M., Shimada, A., Minematsu, T., & Taniguchi, R.-I. (2019). Learning support system for providing page-wise recommendation in e-textbooks. In *Society for Information Technology & Teacher Education International Conference* (pp. 824–831). Association for the Advancement of Computing in Education (AACE).
46. Kamau, J. G., Senaji, T. A., Eng, R., & Nzioki, S. C. (2019). Effect of information technology capability on competitive advantage of the Kenyan banking sector. *International Journal of Technology and Systems, 4*(1), 1–20.
47. Ehikioya, S. A., & Lu, S. (2019). A path analysis model for effective E-commerce transactions. *African Journal of Computing and ICT, 12*, 55–71.
48. Zhang, S., Tay, Y., Yao, L., & Sun, A. (2018). Dynamic intention-aware recommendation with self-attention. *arXiv preprint arXiv:1808.06414.*

Index

© Springer Nature Switzerland AG 2020
S. Paiva, S. Paul (eds.), *Convergence of ICT and Smart Devices for Emerging
Applications*, EAI/Springer Innovations in Communication and Computing,
https://doi.org/10.1007/978-3-030-41368-2

Printed in the United States
by Baker & Taylor Publisher Services